MANAGING ORGANIZATIONAL CHANGE IN NUCLEAR ORGANIZATIONS

The following States are Members of the International Atomic Energy Agency:

AFGHANISTAN
ALBANIA
ALGERIA
ANGOLA
ARGENTINA
ARMENIA
AUSTRALIA
AUSTRIA
AZERBAIJAN
BAHAMAS
BAHRAIN
BANGLADESH
BELARUS
BELGIUM
BELIZE
BENIN
BOLIVIA
BOSNIA AND HERZEGOVINA
BOTSWANA
BRAZIL
BRUNEI DARUSSALAM
BULGARIA
BURKINA FASO
BURUNDI
CAMBODIA
CAMEROON
CANADA
CENTRAL AFRICAN REPUBLIC
CHAD
CHILE
CHINA
COLOMBIA
CONGO
COSTA RICA
CÔTE D'IVOIRE
CROATIA
CUBA
CYPRUS
CZECH REPUBLIC
DEMOCRATIC REPUBLIC OF THE CONGO
DENMARK
DOMINICA
DOMINICAN REPUBLIC
ECUADOR
EGYPT
EL SALVADOR
ERITREA
ESTONIA
ETHIOPIA
FIJI
FINLAND
FRANCE
GABON
GEORGIA
GERMANY
GHANA
GREECE
GUATEMALA
HAITI
HOLY SEE
HONDURAS
HUNGARY
ICELAND
INDIA
INDONESIA
IRAN, ISLAMIC REPUBLIC OF
IRAQ
IRELAND
ISRAEL
ITALY
JAMAICA
JAPAN
JORDAN
KAZAKHSTAN
KENYA
KOREA, REPUBLIC OF
KUWAIT
KYRGYZSTAN
LAO PEOPLE'S DEMOCRATIC REPUBLIC
LATVIA
LEBANON
LESOTHO
LIBERIA
LIBYA
LIECHTENSTEIN
LITHUANIA
LUXEMBOURG
MADAGASCAR
MALAWI
MALAYSIA
MALI
MALTA
MARSHALL ISLANDS
MAURITANIA
MAURITIUS
MEXICO
MONACO
MONGOLIA
MONTENEGRO
MOROCCO
MOZAMBIQUE
MYANMAR
NAMIBIA
NEPAL
NETHERLANDS
NEW ZEALAND
NICARAGUA
NIGER
NIGERIA
NORWAY
OMAN
PAKISTAN
PALAU
PANAMA
PAPUA NEW GUINEA
PARAGUAY
PERU
PHILIPPINES
POLAND
PORTUGAL
QATAR
REPUBLIC OF MOLDOVA
ROMANIA
RUSSIAN FEDERATION
RWANDA
SAN MARINO
SAUDI ARABIA
SENEGAL
SERBIA
SEYCHELLES
SIERRA LEONE
SINGAPORE
SLOVAKIA
SLOVENIA
SOUTH AFRICA
SPAIN
SRI LANKA
SUDAN
SWAZILAND
SWEDEN
SWITZERLAND
SYRIAN ARAB REPUBLIC
TAJIKISTAN
THAILAND
THE FORMER YUGOSLAV REPUBLIC OF MACEDONIA
TOGO
TRINIDAD AND TOBAGO
TUNISIA
TURKEY
UGANDA
UKRAINE
UNITED ARAB EMIRATES
UNITED KINGDOM OF GREAT BRITAIN AND NORTHERN IRELAND
UNITED REPUBLIC OF TANZANIA
UNITED STATES OF AMERICA
URUGUAY
UZBEKISTAN
VENEZUELA
VIET NAM
YEMEN
ZAMBIA
ZIMBABWE

The Agency's Statute was approved on 23 October 1956 by the Conference on the Statute of the IAEA held at United Nations Headquarters, New York; it entered into force on 29 July 1957. The Headquarters of the Agency are situated in Vienna. Its principal objective is "to accelerate and enlarge the contribution of atomic energy to peace, health and prosperity throughout the world".

IAEA NUCLEAR ENERGY SERIES No. NG-T-1.1

MANAGING ORGANIZATIONAL CHANGE IN NUCLEAR ORGANIZATIONS

INTERNATIONAL ATOMIC ENERGY AGENCY
VIENNA, 2014

COPYRIGHT NOTICE

Printed by the IAEA in Austria
April 2014
STI/PUB/1603

IAEA Library Cataloguing in Publication Data

Managing organizational change in nuclear organizations. — Vienna : International Atomic Energy Agency, 2014.
p. ; 30 cm. — (IAEA nuclear energy series, ISSN 1995–7807 ; no. NG-T-1.1)
STI/PUB/1603
ISBN 978–92–0–140910–2
Includes bibliographical references.

1. Nuclear facilities — Management. 2. Nuclear facilities — Safety measures. 3. Organizational change — Planning. I. International Atomic Energy Agency. II. Series.

IAEAL 14–00890

FOREWORD

One of the IAEA's statutory objectives is to "seek to accelerate and enlarge the contribution of atomic energy to peace, health and prosperity throughout the world." One way this objective is achieved is through the publication of a range of technical series. Two of these are the IAEA Nuclear Energy Series and the IAEA Safety Standards Series.

According to Article III.A.6 of the IAEA Statute, the safety standards establish "standards of safety for protection of health and minimization of danger to life and property". The safety standards include the Safety Fundamentals, Safety Requirements and Safety Guides. These standards are written primarily in a regulatory style, and are binding on the IAEA for its own programmes. The principal users are the regulatory bodies in Member States and other national authorities.

The IAEA Nuclear Energy Series comprises reports designed to encourage and assist R&D on, and application of, nuclear energy for peaceful uses. This includes practical examples to be used by owners and operators of utilities in Member States, implementing organizations, academia, and government officials, among others. This information is presented in guides, reports on technology status and advances, and best practices for peaceful uses of nuclear energy based on inputs from international experts. The IAEA Nuclear Energy Series complements the IAEA Safety Standards Series.

In 2001, the IAEA published Managing Organizational Change (IAEA-TECDOC-1226). It was written at a time of great changes within the industry, highlighting the need to effectively manage these changes. That publication provided a description of the basic principles for managing change in nuclear utilities based on the practices being used in many Member States at that time. These practices were being utilized by both the licensee of the utility and regulators to monitor and implement effective change while remaining focused on safe and reliable operation. The present publication has been written to update and refine IAEA-TECDOC-1226.

Organizations engaged in applying nuclear technology have undergone many changes since their inception, but the speed and nature of the changes seem to be increasing. Organizations have always faced technical challenges in exploiting the use of nuclear power safely and effectively; while these challenges remain, there are other emerging challenges related to the need for organizational change to adapt to the modern energy supply market. Organizational change inevitably involves people, and the human elements of change can be as demanding as some of the historical technical challenges faced by the nuclear industry.

Much has been learned since the publication of IAEA-TECDOC-1226 about planning change, about human behaviour during change and about implementing organizational changes. Successful implementation of organizational change requires: an understanding of the influence of organizational culture; strong leadership; involvement of the workforce throughout the change process; good communication; and effective regulation. In particular, it is necessary for both facilities and their regulators to understand that safety can be preserved, and even enhanced, in successful change, provided both parties are aware of potential safety issues during the change period, when vulnerabilities may be exposed.

This publication has been developed for all levels of management involved in implementing changes within their areas of responsibility. It should be of particular interest to managers who have a lead role in implementing organizational change. The purpose of this publication is not to define when change should be pursued within an organization, or when such change should be avoided. Rather, it provides guidance to organizations on how to respond to the inevitability of change and on how to progress in a manner that enhances the likelihood of the successful implementation of organizational change.

The safety culture that has proved successful in today's nuclear facilities has developed over many years. Many nuclear facilities have difficulty sustaining this culture during the transitions that are an intrinsic part of change. Properly managed, however, change can enhance nuclear safety, reliability and cost competitiveness. In recent years, many facilities have observed that the most successful and sustained changes have come about when individuals at all levels have been involved in establishing the goals for change, planning changes and implementing them. This emphasizes the need for wide participation in the organizational change effort.

This publication is based on the experience of Member States and identifies important factors to consider when undertaking organizational changes in a nuclear facility. The most important source of information for Member States was a series of workshops, organized through the IAEA's technical cooperation programme, which provided a forum for exchanging experience on the effective management of change. The main focus was on common difficulties, possible solutions and good practices to improve overall performance, with due regard to safety. The results and considerations of the workshops were utilized in the preparation of this publication.

The IAEA wishes to thank the senior management of those nuclear organizations that provided relevant information and source materials. Special thanks go to experts from the Czech Republic, Finland, Hungary, Italy, Japan, Slovakia, Spain, the United Kingdom and the United States of America who contributed to the preparation of this publication. The IAEA officers responsible for this publication were J.P. Boogaard, K. Dahlgren and J. Majola of the Division of Nuclear Power.

CONTENTS

1. INTRODUCTION

1.1. BACKGROUND

It is widely recognized that engineering changes, if not properly considered and controlled, can have potentially major safety implications. This is particularly true for the nuclear industry, where considerable efforts are made to establish and maintain safety cases for nuclear facilities. It can also be an issue for other apparently less hazardous operations, where full safety cases do not need to be developed. The requirement for risk assessment is aimed at addressing these issues, and especially potential safety concerns.

What has often not been so widely recognized is that organizational changes can also have a major impact on safety and effectiveness and, in some cases, can have consequences as serious as failures in engineering control. It is important to have a management system that ensures that people related changes do not inadvertently have an impact on safety and performance.

A management system based on the requirements stated in IAEA Safety Standards Series No. GS-R-3 (hereinafter referred to as IAEA GS-R-3) [1] and supporting guidance, such as that in IAEA Safety Standards Series Nos GS-G-3.1 [2] and GS-G-3.5 [3], and in IAEA-TECDOC-1226 [4] and IAEA-TECDOC-1491 [5], would be such a system. This system requires the organization to evaluate organizational changes and classify them according to their importance to safety. It requires the organization to justify each organizational change and to plan, control, communicate, monitor, track and record the implementation of each change to ensure that safety is not compromised.

Many of the potentially adverse impacts on safety of organizational changes can be avoided if consideration is given to the effects of changes before they take place. Both the final organizational arrangements resulting from the implementation of the change and the transitional arrangements need to be considered. A management system based on IAEA GS-R-3 can impose such a discipline within an organization.

Human and organizational factors have a clear influence on nuclear safety and economy. There are many examples where misdirected changes have had a large impact, either as an incident or simply as lost production. If managers in the nuclear industry obtain a better understanding of organizational change and its various dynamics, this knowledge will serve them and their industry well in meeting future challenges.

Historically, the nuclear industry has taken a risk averse approach in its operations and activities and continues to give priority to the avoidance or minimization of risk. However, introducing any change in a nuclear facility has the potential to have an impact on safety; therefore, it is essential to have processes that support the effective management of change. These processes enable the facility to identify the risks associated with a change and to manage the risks properly while realizing the benefits of the intended change. The availability and use of change processes by employees can be regarded as tangible evidence of their commitment to a strong safety culture. Management systems based on IAEA GS-R-3 require organizations to establish, implement, assess and continually improve such processes.

Typically, the concept of organizational change involves organization wide change, as opposed to smaller changes such as adding a new person or minor alterations to a programme. Examples of organization wide change include a change in mission, restructuring, business process re-engineering, outsourcing of functions, mergers and major collaborations, use of new technologies, etc. The management of design changes or the management of continual improvements and process changes, while important, is not the focus of this publication.

In order for an organizational change effort to be successful, two levels of change must be addressed: the strategic level and the operational level. The differences between strategic and operational change lie in the scope of effort, the people most involved and the outcome goals. Leadership for strategic change comes from the senior levels of an organization. Small teams of selected individuals, perhaps including consultants, are often involved. The primary goals of the strategic change phase of organizational change are the generation of recommendations and the establishment of momentum for change. Operational change is the effort that drives change deep into an organization. It focuses on implementation at the local level within an organization. Operational change involves many more people, with leadership coming from senior managers, middle management and frontline supervisors. The primary goal of operational change is to implement and sustain desired changes. Effectively managing both the strategic and operational levels of change can mean the difference between success and failure.

IAEA-TECDOC-1226 [4] was published in 2001 to provide senior management with principles and practical guidance for managing change in nuclear facilities, based on practices used in many Member States to implement effective change while keeping the focus on safe and reliable operation of nuclear power plants. This manual supersedes that publication, featuring a different title to reflect the broad applicability of its content to all facilities involved in the peaceful use of nuclear technology, not just nuclear power plants.

This publication was initiated and largely written before the events at the Fukushima Daiichi nuclear power plant in 2011. Many reports and official briefings are now available, which include lessons learned and other appropriate material, based on preliminary assessment results. It is recommended that users of this publication consider these when developing or implementing activities aimed at the management of change, and evaluate the safety impact or any additional requirements that may apply to their organization's operations using the lessons learned from the Fukushima Daiichi accident.

1.2. OBJECTIVE

The objective of this publication is to assist the management of nuclear facilities in identifying, planning and implementing organizational change. The driving force for the change may be internal or external.

This publication makes the basic assumption that any change made within a facility applying nuclear technology has the potential to impact safety and effectiveness. Therefore, it provides a description of the basic principles to manage and implement organizational changes to the facility effectively while remaining focused on safe and reliable operation.

In preparing this publication, efforts have been made to ensure consistency with IAEA safety standards, comprising the Safety Fundamentals, Safety Requirements and Safety Guides, particularly as they apply to the management system for facilities and activities [1–3]. Specifically, this publication can be used to help establish processes for managing organizational change required in management systems based on, or meeting, the requirements of IAEA GS-R-3 [1].

1.3. SCOPE

The guidance contained in this publication is relevant to all organizational changes within facilities, whether they are related to the structure of the facility itself or to the people. Accordingly, the process described herein can be applied to changes of varying types and magnitudes.

The methods and processes described in this publication are based on practical experience within Member States and good practices specified in organizational change literature.

The primary intended users of this publication are:

— Managers at all levels who are developing and implementing organizational changes within their respective areas of responsibility;
— Regulators involved in assessing organizational changes in nuclear facilities to ensure that the changes are properly planned and executed so that safety and reliability are maintained and possibly enhanced.

1.4. STRUCTURE

Following this introduction, Section 2 outlines a basic process for managing organizational change. It deals with planning, implementing, embedding and sustaining organizational changes. Sections 3–5 highlight or provide supporting and more detailed information on some of the aspects or the major phases of the basic change process described in Section 2. Sections 6 and 7 describe the more practical considerations needed to enact a change. They provide further insights when applying or using the basic process for managing organizational change.

In particular, Section 3 focuses on identifying and introducing the need for change. It outlines the major causes or drivers of organizational changes, the impacts on safety that can occur during organizational change, and the importance of anticipating organizational change.

Section 4 deals with harnessing support for change. It discusses the role of leadership, organizational culture and regulators in successfully implementing change at nuclear facilities. It also discusses the importance of communication during organizational change.

Section 5 deals with building organizational structures and human resource competencies to support and effectively implement organizational change. It addresses the importance of a well designed organizational structure and discusses the role of teams and the role of training and coaching in organizational change.

Section 6 deals with the safety impact and implications that could result from the ineffective management of change. It discusses the necessity for any organization to ensure that the safety impact possibilities are addressed during any change process.

Section 7 describes the practical steps needed to prepare for and implement organizational changes effectively. This includes process preparations plus cultural issues. In practical terms, this section is a 'how to' guide to the management of change, but it is essential that the reader first understand the requirements and implications of the earlier sections before using it to implement change.

Section 8 describes the relationship between the steps to implement a change and theoretical aspects. Suggested theoretical references are given for each of these steps.

The annexes, on the companion CD-ROM, provide supporting information on the leadership and levels of change, on rewards and recognition (that can be an important factor in ensuring that organizational change is embedded), and an example of a questionnaire that can be used to evaluate the degree of readiness of organizations for change.

A list of references and a bibliography provide information on the sources used in this publication and for further reading.

2. MANAGING ORGANIZATIONAL CHANGE

The IAEA Safety Requirements (namely, IAEA Safety Standards Series No. GS-R-3) identify managing organizational change as one of the key generic management system processes that should be developed in the management system. There are two main requirements:

- "5.28. Organizational changes shall be evaluated and classified according to their importance to safety and each change shall be justified.
- "5.29. The implementation of such changes shall be planned, controlled, communicated, monitored, tracked and recorded to ensure that safety is not compromised." (Ref. [1], p. 14.)

The IAEA has also published a Safety Guide [2] to support its Safety Requirements, and provides detailed guidance on managing organizational change. It states that when organizational change is necessary, no reduction in the level of safety achieved should be acceptable, even for short periods of time, without appropriate justification and approval. When major organizational changes are planned, they should be rigorously and independently scrutinized. Senior management should remain aware that it has the ultimate responsibility for safety and should ensure that safety considerations are given a priority commensurate with their significance during any process of major change. Individuals should be made aware of how their responsibilities will change both during and after organizational changes. Senior management should develop a specific process to manage and review organizational changes.

Because organizations are open systems — dependent on their environments — and because the environment does not stand still, organizations must develop internal mechanisms to facilitate planned change. Change efforts that are planned — proactive and purposeful — are what is meant by managing change.

When addressing the changes, the organization needs to deploy appropriate resources that will carry on its strategy and implement the change(s). Depending on the type of change, the approach to be applied will be different. This 'graded' approach is consistent with para. 2.6 in IAEA GS-R-3 [1].

To determine what type of change needs to be applied, it is important to identify:

— What is impacted/what is to be changed;
— The type of change;
— The extent of the change (e.g. the depth of the impact);
— The level of the change (grading);
— The safety impact of the change.

These factors will be analysed in the next sections. One of the ways of "thinking, seeing and acting for powerful, purposeful change in organizations" [6] is through 'appreciative inquiry', which is introduced in Section I–3 of Annex I.

2.1. MODEL FOR MANAGING ORGANIZATIONAL CHANGE

Figure 1 represents a model for organizational change. It can be broken down into a series of steps. Certain forces initiate change. A change agent initiates the change and chooses the intervention action. Implementation of the intervention contains two parts: what is done and how it is done. The 'what is done' stage requires three phases: unfreezing the status quo, movement to a new state and refreezing the new state to make it permanent. The 'how' stage refers to the tactics used by the change agent to implement the change process. The change, if successful, improves organizational effectiveness. Changes do not take place in a vacuum. A change in one area of the organization is likely to initiate new forces for other changes. The feedback loop in Fig. 1 acknowledges that this model is dynamic.

2.2. POTENTIAL IMPACT ON SAFETY OF ORGANIZATIONAL CHANGE

Safety should be treated as a priority commensurate with its significance during any organizational change. This is particularly important in the nuclear industry, which has historically understood the critical importance of safety. However, all aspects of safety, nuclear or occupational, and other requirements for managing an organization that have an impact on safety, including environmental, financial, quality, and security requirements, must be considered during the change process.

An organization should objectively review the state of its safety culture to ascertain whether there are any indications that it may be weakening before embarking on organizational change. IAEA GS-G-3.5 [3] describes how to achieve the attributes of a strong safety culture.

2.3. FORCES INITIATING CHANGE

The rate of organizational change has not slowed in recent years, and may even be increasing. The rapid and continual innovation in technology is driving changes to organizational systems and processes. For example, the Internet is enabling much faster and easier access to knowledge. Add to this the increased expectation of employees as they move more freely between organizations. The global nature of business has resulted in the removal of previous international market barriers. Constant change has become a fact of organizational life. Change affects all organizations but, just as it manifests itself in various ways, there are very diverse motives for bringing it about. Every instance of change is unique because every business has different needs and imperatives.

How does an organization know that change is necessary? It may be the identification of an opportunity upon which management wants to capitalize. More often, however, it is in anticipation of, or in reaction to, a problem. These opportunities and problems may exist inside the organization, outside the organization, or both. Therefore, the forces for change can be divided into two categories: external and internal. External forces for change are factors arising outside the organization that is to be changed. Internal forces for change, on the other hand, come into play where changes are initiated by the business itself, essentially as a result of a desire to evolve and improve. External forces are by far the most common. Some of the common causes of organizational change are as follows:

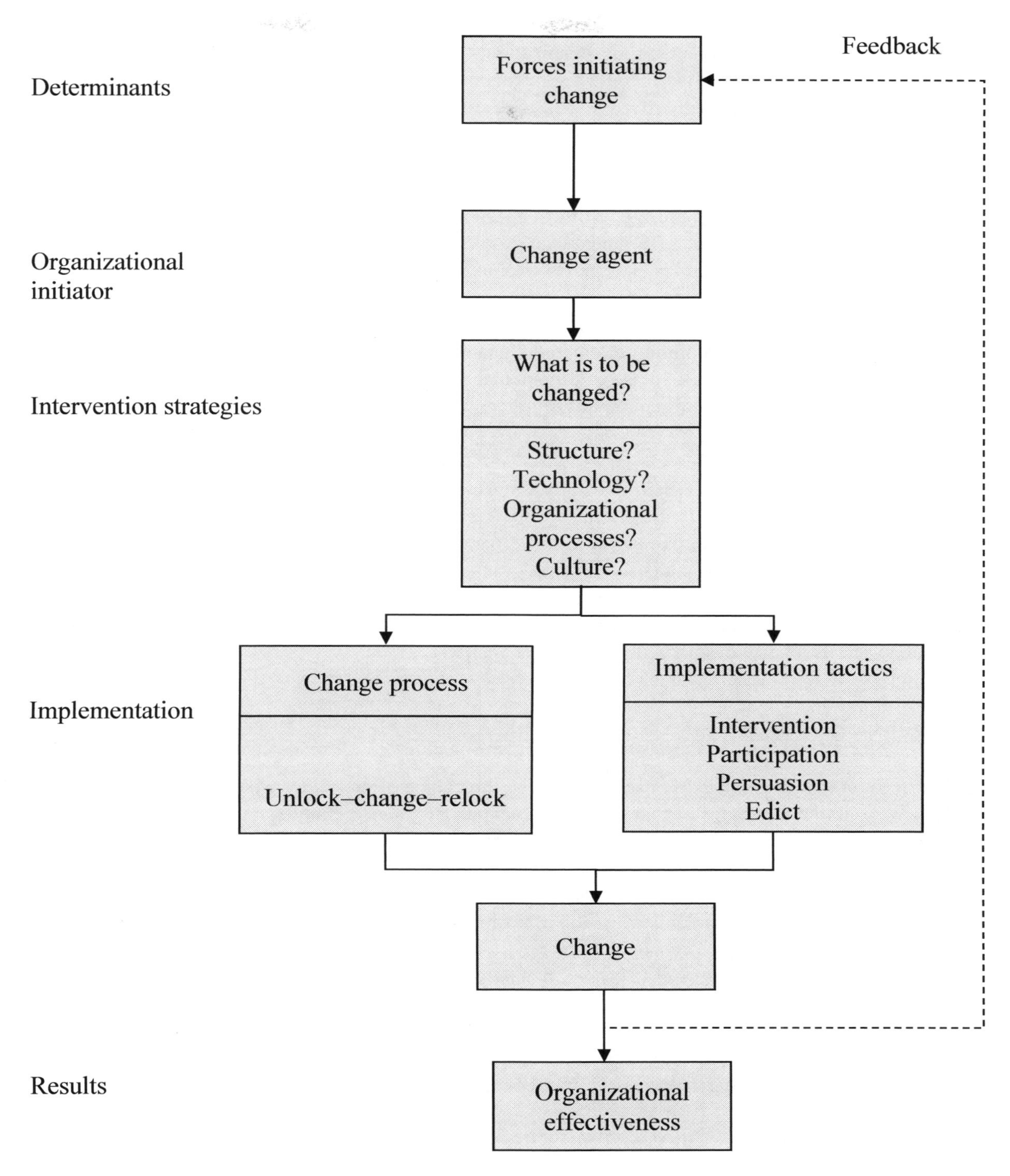

FIG. 1. Model for managing organizational change.

- Strategic refocus: When an organization changes its business processes to adopt a new strategy, organizational change ensues.
- Structural change: When new administrative processes are introduced, organizational change results. Old processes are replaced by new ones and employees retrain to operate the new systems.
- Technological innovation: This is an increasingly significant force for change and is one that seems to be constantly increasing.
- Mergers and acquisitions: These are one of the most frequent causes of organizational change, often resulting in a change in organizational culture.

— Change in market: As the market for an organization's product or service reaches maturity, market growth and profits begin to diminish. This encourages the organization to reposition itself or exit the market, and this can result in organizational change.
— Regulatory change: There may be changes in the regulatory requirements resulting from national or international directives.
— Downsizing: To survive, many organizations adopt new ways of working using modern technology with less need for people. Organizational change results in a leaner organization.
— Political causes: A change in government can result in a new regulatory environment that forces an organization to make significant changes in its systems and processes.

Effective change management incorporates the anticipation of change by frequent or continual monitoring of sources of change, and the incorporation of emergent change triggers into organizational planning. Thus, some organizations practice what is referred to as 'environmental scanning' — a process of continually reviewing the business, societal and governmental influences on the organization.

2.4. ORGANIZATIONAL INITIATOR

Change agents initiate change in an organization. Change agents are generally those in power and typically include senior managers, internal specialists, managers of major organizational units, and occasionally consultants brought in from the outside. The outside consultant brings to the organization objectivity to analyse the problems and the expertise to offer valuable suggestions for change. Change agents are the intermediaries between the forces initiating change and the choice of intervention strategy.

2.5. INTERVENTION STRATEGIES

The term 'intervention strategy' is used to describe the choice of means by which the change process takes place. Strategies tend to fall into one of four categories: people, structure, technology and organizational processes.

The people classification refers to interventions that attempt to change the behaviour of employees, their attitude or values and can often entail the challenging task of changing the organizational culture.

The structure classification includes changes affecting the organizational hierarchy, allocation of rewards, degree of formalization, and addition or elimination of positions, departments or divisions.

The technology classification encompasses modifications in the equipment used by employees, the interdependencies of work activities, and changes that affect the interrelationships between employees and the technical demands of their jobs.

The organizational processes classification considers changing processes used by the organization to carry out its business, be they administrative processes such as those for decision making and communication, or core operational processes for delivering the products (items or services) of the organization, and their support processes (e.g. information technology).

The type of intervention is strictly dependent on the type of change needed.

2.5.1. Types of change

Typically, the expression 'organizational change' refers to a significant change in the organization, such as a reorganization or adding a new product or service. This is in contrast to smaller changes, such as adopting a new procedure. It is helpful to think of organizational change in terms of various types as described below:

— *Organization wide versus subsystem change.* Examples of organization wide change might be a major restructuring, merger or 'right-sizing'. Cultural change is another example of an organization wide change. Examples of a change in a subsystem might include the addition or removal of a product or service, reorganization of a certain department, or implementation of a new process to deliver products or services.

— *Transformational versus incremental change.* An example of transformational (or radical, fundamental) change might be changing an organization's structure and culture from a traditional top-down, hierarchical structure to one comprising self-directing teams. Another example might be business process re-engineering, which tries to take apart (at least on paper, at first) the major parts and processes of the organization and then put them back together in a more optimal manner. Examples of incremental change might include continuous improvement as a quality management process or implementation of a new computer system to increase efficiency.

— *Remedial versus developmental change.* Change can be made to remedy current situations, for example, to improve the poor performance of a service or help the organization become more proactive and less reactive, or to address large budget deficits. Remedial projects often seem more focused and urgent because they are addressing a current major problem. It is often easier to determine the success of these projects because the problem is either solved or not. Change can also be developmental — to make a successful situation even more successful, for example, to expand the number of customers served, or to duplicate successful products or services. Developmental projects can seem more general and less well defined than remedial projects, depending on how specific are the goals and how important it is for members of the organization to achieve those goals. Sometimes, if a developmental change is not made, the need for a remedial change may arise. Also, organizations may recognize current remedial issues and then establish a development vision of change to address the issue.

— *Unplanned versus planned change.* Unplanned change usually occurs because of a sudden major surprise to the organization, which causes its members to respond in a highly reactive and possibly disorganized way. Unplanned change might occur when the chief executive suddenly leaves the organization, when a significant public relations problem arises or when other disruptive situation arises. Planned change occurs when leaders in the organization recognize the need for a major change and proactively prepare a plan to accomplish the change. Planned change, even though based on a proactive and well considered plan, often does not occur in a highly organized way.

There is another way of considering the different types of change before embarking on an organizational change project that supplements the above. The different types of change can be categorized in terms of three main variables:

— Depth of change;
— Speed of change;
— How change is implemented.

Depth of change is the degree to which the change affects the nature of the organization's business, which may be altered profoundly or slightly, or to some degree in between.

Speed of change varies considerably, not only from one organization's business to another, but also within the same business. It is at its greatest when change is dictated by circumstances, when managers are aware of the need for it and when the organization itself is well adapted to it. If all these conditions are met, even profound changes can be made in a short time. The more accustomed an organization is to changing, the quicker it becomes at doing so. If change is only experienced occasionally and is regarded as a major ordeal, it will take a great deal of time to overcome resistance and motivate people to support the effort and implement the necessary changes.

The way change is implemented can vary greatly; it may be forcibly imposed or it may be the result of total consensus. Change can be imposed in situations where there are opposing interests and common ground cannot be found. Imposed change generally takes place in an organization with a strong hierarchy where authority is accepted. Nevertheless, it is extremely unusual for the entire change process to be imposed, since its success actually depends on the cooperation of those involved. Change by consensus, on the other hand, is characterized from the outset by the complete support of all concerned. Although it can be difficult to obtain such support, the aim is to improve the chances of success by a high initial level of motivation and participation.

2.5.2. Conditions for the effective management of change

To manage change effectively, organizational leaders need to establish the following:

- — A clear understanding of why the change is necessary;
- — A vision of what the organization should look like after the change, and a direction towards that vision;
- — A clear sense of the organization's purpose;
- — A clear sense of the organization's interdependence with its external environment;
- — Clarity about achievable scenarios or descriptions of possible end states;
- — Effective organizational structures to manage the types of work required;
- — Effective use of advanced technology;
- — Good communications delivering coherent and transparent information that encourages the involvement of people;
- — Reward systems that equally reflect organizational priorities and the needs of individuals for dignity and growth.

To establish the above, executives and managers need both skills and understanding in areas that were not a priority in the past. They must have a good grasp of: the changing nature of work in the information age; telecommunications technology and its potential role for the organization; the nature of culture and what it takes to change it; the significant role of values in an organization's life; and finally the concepts of managing effective change and of balancing stability and change.

2.5.3. Characteristics of a strong safety culture

The five key characteristics that, through their attributes are essential to achieve a strong safety culture, are defined in Ref. [3] and, for ease of use, are listed below:

- — Safety is a clearly recognized value;
- — Leadership for safety is clear;
- — Accountability for safety is clear;
- — Safety is integrated into all activities;
- — Safety is learning driven.

Not all characteristics of a positive safety culture develop at the same rate; some characteristics will be more challenging than others to develop. However, the absence of many of the attributes should be a cause for concern, as this would indicate that the organization's safety state is vulnerable to destabilization. A planned organizational change effort could be the trigger that catalysed such a destabilization.

Where many of the above characteristics of a strong safety culture are absent the organization should seriously consider postponing any major organizational change, until it has strengthened its safety culture by eliminating or mitigating the identified weaknesses. Or it should devise an organizational change strategy that also incorporates measures to strengthen its safety culture.

The weaknesses may have to be dealt with in a phased manner and, in some cases, may involve a series of minor organizational changes. Provided these are addressed systematically, avoiding dealing with too many in parallel, the organization's safety culture can be stabilized prior to, or as part of, embarking on a major organizational change effort.

More information related to the leadership of change is provided in Annex I.

2.5.4. Factors that determine success

The successful implementation of organizational change is challenging and in many cases can be a disappointing failure. However, organizations armed with the knowledge of factors that have the biggest impact on their success can take steps to influence or change those factors. The five factors that play a critical role in the successful implementation of change — threat, commitment, plans, progress and strategy — are discussed below:

(1) *Threat.* Of the five factors that impact the probability of a successful change implementation, threat is the most powerful. The presence of a threat seems to make it far more likely that the change will be implemented. It is easier to motivate employees to change when the survival of the organization is threatened.

(2) *Commitment.* The second most important factor affecting the success of a change effort is the degree of commitment that employees have to the change. Although commitment is strongly linked to threat, the presence of a threat is not required for commitment. Commitment to change is not a yes or no issue and it is best understood in the context of a scale that ranges from 0 to 100% as shown in Table 1.
The employee at the 100% level is committed to the change both intellectually and emotionally — in other words, in thought and deed. Those at the 75% level have a strong intellectual commitment to the change. They know that the change will help the organization become more successful. However, the person at this level is not committed emotionally and may have difficulty aligning day to day behaviour with that required in the change effort. In a stressful situation, the 75% committed person may revert to old behaviour patterns. At the 50% level, a person shows real compliance with the change effort, but no real commitment on either an intellectual or emotional level. This person goes along with the change effort because it is expected of him or her. People at the 25% level support the change effort in public situations, but in private they are cynical. They may view the change effort as another short lived management initiative that will not be fulfilled. The person at the 0% level is openly non-compliant and believes that the change is a waste of time and voices this belief.

(3) *Plans.* Of the five factors that affect the probability of success, this factor has the least impact. The assessment of current strengths and weaknesses and the development of a plan for a change effort to address them are less important than other factors because a plan, by itself, does little to guarantee a long term commitment. Organizations that have three to five year plans are more likely to sustain the change effort than those operating on a shorter timescale. The plan should not be too aggressive as many organizations make the mistake of trying to do too much too quickly. The plan should set goals that focus on results, and not just on activities that will lead to the results.

(4) *Progress.* This factor impacts on the probability of success, i.e. the progress made in the implementation of the change effort. The further along an organization is with the change effort, the less likely the effort will be dropped. It is only when the change is fully integrated into the organization and operating successfully on a day to day basis that progress can be regarded as complete.

(5) *Strategy.* This factor influences the success of the effort, specifically how the change strategy is implemented. For example, organizations that create a hierarchy of committees to oversee implementation employ a strategy that often results in failure. Another key feature of strategy is its deployment across the organization. To be effective, the strategy must involve all functions and levels of the organization.

TABLE 1. SCALE OF COMMITMENT

100%	75%	50%	25%	0%
Commitment on both the intellectual and emotional levels	Strong intellectual commitment; little emotional commitment	Compliance only	Lip service	No compliance or commitment

2.6. IMPLEMENTATION

2.6.1. Change process

Successful change requires 'unlocking' the status quo, moving to a new state, and relocking the change to make it permanent. Implicit in this three stage change process is the recognition that the mere introduction of change does not ensure the elimination of the pre-change condition or the fact that the change will prove enduring.

The status quo can be considered an equilibrium state. To move from this equilibrium — to overcome the pressures of both individual resistance and group conformity — unlocking is necessary. This can be achieved in one of three ways. The driving forces, which direct behaviour away from the status quo, can be increased. The restraining forces, which hinder movement from the existing equilibrium, can be decreased. A third choice is to combine the first two approaches.

Once unlocking has been accomplished, the change can be implemented. This is where the change agent introduces one or more intervention strategies. In reality, there is no clear line separating unlocking and changing. Many of the efforts made to unlock the status quo may, of themselves, introduce change. So the tactics that the change agent uses for dealing with resistance may work on unlocking and/or moving.

Assuming that the change has been implemented, if it is to be successful the new situation needs to be 'relocked' so that it can be sustained over time. Unless this last step is attended to, there is a very high likelihood that the change will be short-lived and employees will attempt to revert to the prior equilibrium state. The objective of relocking is to stabilize the new situation by balancing the driving and restraining forces. How is relocking done? Basically, it requires systematic replacement of the temporary forces with permanent ones. It may mean formalizing the driving or restraining forces. The formal rules and regulations governing the behaviour of those affected by the change should be revised to reinforce the new situation. Over time, the group's norms will evolve to sustain the new equilibrium. But until that point is reached, the change agent will have to rely on more formal mechanisms.

There are several factors that determine the degree to which a change will become permanent. The reward allocation system is critical and if rewards fall short of expectations over time, the change is likely to be short lived. During an organizational change effort, the reward and recognition system should only be finalized after significant progress has been made with the following:

— Development of an adaptive organizational culture that fosters trust and receptiveness to change;
— Improvement of leadership and management skills;
— Development of a teambuilding structure;
— Development of a well designed organizational structure.

General information on providing recognition and reward is provided in Annex I. If a change is to be sustained, it needs the support of a sponsor. This individual, typically high in the management hierarchy, provides legitimacy for the change. Once senponsorship is withdrawn from a change project, there are strong pressures to return to the old equilibrium state.

A failure to communicate information on expectations can reduce the degree of sustainability of the change, as people need to know what is expected of them as a result of the change.

Group force is another important factor. As employees become aware that others in their group accept and sanction the change, they become more comfortable with it.

Commitment to the change should lead to a greater acceptance and permanence. As noted earlier, if employees participate in the change decision, they can be expected to be more committed to ensuring that it is successful.

Change is less likely to become permanent if it is implemented in a single unit of an organization. Therefore, the more diffusion in the change effort, the more units will be affected and the greater legitimacy the effort will carry.

The above factors remind us that the organization is a system and that planned change will be more successful when all the parts within the system support the change effort. Successful change also requires careful balancing of the system. All changes, regardless of how small, will have an impact outside the area in which they were implemented. No change can take place in a vacuum.

2.7. IMPLEMENTATION TACTICS

Parallel to the change process in the implementation stage is the decision on what tactics should be used to introduce the planned change. The four tactics [7] that have generally been used by change agents are as follows:

(i) *Intervention.* The change agents sell the change rationale to those who would be affected. They argue that current performance is inadequate and establish new standards. The agents refer to comparable organizations with better performance to justify the need for change, and then describe how current practices can be improved. The change agents may form a task force made up of affected personnel. Change agents retain the power to veto any of the task force's recommendations.

(ii) *Participation.* By adopting this tactic, the change agents delegate the implementation decision to those who will be affected. The change agents stipulate the need for change, create a task force to do the job, assign members to the task force, and then delegate authority for the change process to the task force with a statement of expectations and constraints. Change agents who use this tactic give full responsibility to the task force for implementation and exercise no veto over its decision.

(iii) *Persuasion.* The change agents identify the opportunity for change, but then take a relatively passive role by inviting interested internal staff or outside experts to present their ideas for bringing about change. Change agents only become active after various ideas have been presented. Those who will be affected choose the best ideas for implementing the change.

(iv) *Edict.* When this tactic is used, change agents merely announce changes and use memos and formal presentations to convey their decision.

In practice, how popular and successful is each of these implementation tactics? A study of 91 cases found persuasion to be most widely used, occurring in 42% of the cases. Edicts were the next most popular, with 23%, followed by intervention and participation, with each slightly less than 20%.

Research demonstrates that there are significant differences in their success rate. The success or otherwise was the retrospective view of the organization that attempted the organizational change. Change directives by managerial fiat are clearly inferior to other options. Edicts were successful in just 43% of the cases. Participation and persuasion achieved success rates of 84% and 73%, respectively. Intervention, while used in less than 20% of the cases, attained a perfect 100% success rate.

2.7.1. Communication to help lower resistance

Possibly the most important pitfall to any change process is not understanding resistance to change. Resistance is likely to be encountered at all levels of the organization. Lack of understanding of it often results at best in frustration and at worst in dysfunctional behaviour, that is, acting against the change, the initiators of the change and the organization itself. Understanding the reasons for resistance and working with it rather than against it will aid greatly in creating a smoother process of change. Also, understanding resistance will help to develop a well thought out communications plan.

Figure 2 shows the resistance pyramid, a framework for understanding the reasons why people resist change. Like Maslow's hierarchy of needs, the resistance pyramid is a succession of levels, in this case resistance levels. Satisfaction at each level reduces resistance at the next level. For example, when we respond to people's need to know, they become more open to learning the new skills and abilities involved in changing. And once they have the new skills, they will gain the confidence to overcome unwillingness to change.

Based on the resistance pyramid concept, what people need first is knowledge. Knowledge can be provided through information about the change process. The information should be based upon what management and employees want to know. People usually want the basic questions answered: What is happening? Why are we doing this? How will it take place? When will it happen? And whom will it affect? Answering these questions for people at each stage of the change process will help them move up to the next level of the pyramid.

The second level of the pyramid — ability — is addressed through training and education. In order to change, people are likely to need new skills. New skills for employees may include operating new equipment or systems, working in teams rather than as individuals, or following revised procedures. Management often needs new skills to create teams and foster teamwork, coach employees to provide them with new skills and apply new procedures. Because ability has a profound impact on the willingness that people have to undertake new activities and make changes, training becomes an integral part of communication and the change process.

The top level of the pyramid is willingness. The acquisition of knowledge at the lowest level and new skills at the middle level will help people to become more willing to change. However, other factors should be addressed as well. The involvement of senior management in the communications process will send signals to the organization about the priority to change. In addition, the individual benefits of change are communicated. Willingness can be increased by several specific actions: (i) establishing individual and team performance goals aligned with the changes to take place; (ii) measuring people against the goals; (iii) establishing effective two way coaching and feedback mechanisms; and (iv) rewarding and recognizing people for achieving goals and implementing the changes.

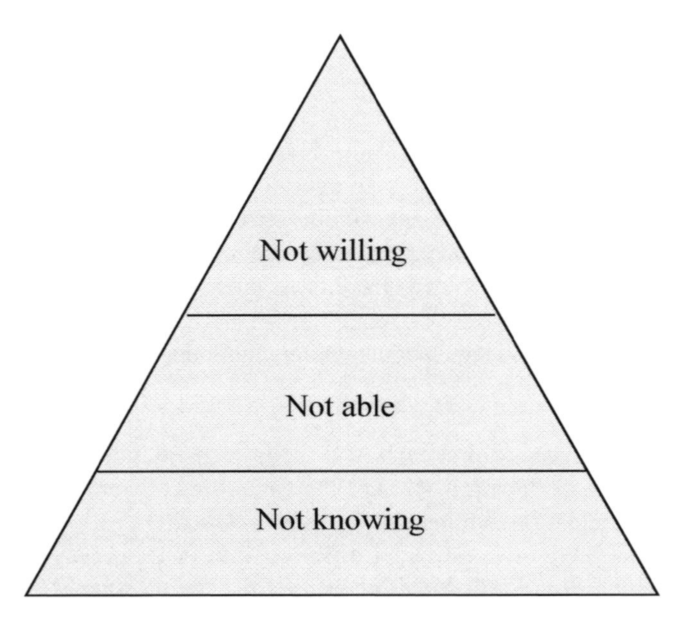

FIG. 2. The resistance pyramid.

Figure 3 shows examples of actions to take in order to move people up the resistance pyramid.

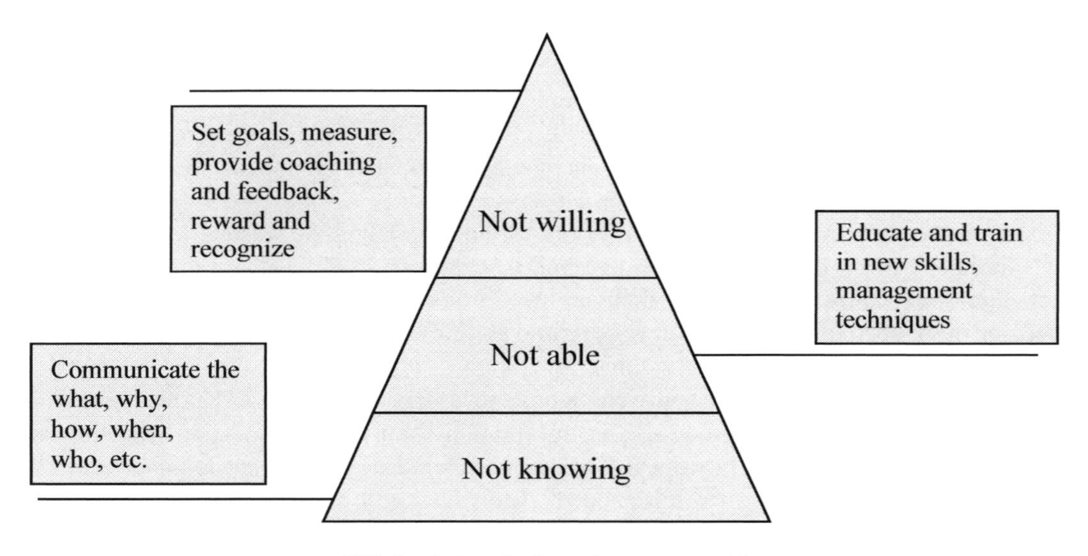

FIG. 3. Actions in the resistance pyramid.

The model in Fig. 2 culminates with change taking place and a resulting impact on organizational effectiveness. Whether that effect is positive, negative, temporary or permanent depends on each of the earlier steps.

Regardless of the outcome, the model shown in Fig. 2 is dynamic. The need for change is continuous, hence the need for the feedback loop. The change model is never at rest.

Examples of tactics that managers or change agents can use for dealing with resistance to change are the following:

— *Education and communication.* By communicating with employees to help them see the logic of the change, resistance can be reduced.

— *Participation.* It is more difficult for individuals to resist a change decision in which they have participated. Assuming that the participants have the expertise to make a useful contribution, their involvement can reduce resistance, obtain commitment and increase the quality of the change decision.
— *Facilitation and support.* Change agents can offer a range of supportive efforts to reduce resistance. When employees' anxieties are high, new skills training and counselling may facilitate adjustment.
— *Negotiation.* This tactic requires the exchange of something of value for a lessening of resistance. Specific reward packages can be negotiated.
— *Co-option.* This tactic seeks to neutralize the resistance of certain individuals by giving them a key role in the change decision.
— *Coercion.* This tactic is the application of direct threats on the resistors and generally involves some form of potential financial penalty.
— *Coaching.* This involves coaching for new skills or application of knowledge.

2.7.2. Why is organizational change difficult to accomplish?

Typically, there are strong resistances to change. People are afraid of the unknown and many may think that things are already just fine. Many doubt that there are effective means to accomplish major organizational change. Organizational change often goes against the very values held by people in the organization. This is why much of the organizational change literature discusses needed changes in the culture of the organization, including changes in people's values and beliefs and the way they enact these values and beliefs.

2.7.3. Why do organizational changes fail?

The following are some of the common reasons why organizational changes do not succeed:

— *Starting with bad advice.* When a change is hastily implemented or attempted without sufficient commitment, it can destroy the credibility of the organization's leaders.
— *Making change an option.* When leaders commit to a change, the message must be that change is not an option. Whenever people have the option not to change, they will not.
— *Focusing only on the process.* Leaders can become so absorbed in the process that they do not notice that no tangible results are being achieved. The activity becomes more important than the result.
— *Focusing only on the results.* This stems from a belief that the end justifies any means. Organizations ignore the human pain of change. It is this insensitivity to people's feelings that not only prevents the change but destroys morale and loyalty in the process.
— *Not involving those expected to implement the change.* A great deal of resentment is aroused when management announces a change and then mandates the specifics of implementation. Employees need to be involved in two ways. First, their inputs and suggestions should be solicited when planning the change. Second, after a change has been committed to, they should be involved in determining the means.
— *Delegating to 'outsiders'.* Although outsiders such as consultants might provide valuable ideas and input, people inside the organization must accept responsibility for the change.
— *Not changing the reward system.* Make sure that reward, recognition and compensation are adjusted for the desired change.
— *Lacking leadership commitment.* For change to happen, everyone involved must buy in. Leaders must take the first steps. Change grinds to a halt when the leaders do not demonstrate the same commitment they expect from others.
— *Not following through.* The best planning is worthless if not implemented and monitored for progress. Responsibility must be clearly defined for making sure that follow-through is timely.

The validity of the above comments is reinforced by observations, which are described in Ref. [8].

In studying organizational change over several decades, it has been noted that whenever human communities are forced to adjust to shifting conditions, there are difficulties. Some of the most common errors when transforming an organization are:

— Allowing too much complacency;
— Failing to create a sufficiently powerful guiding coalition;
— Underestimating the power of vision;
— Under-communicating the vision;
— Permitting obstacles to block the new vision;
— Failing to create short term wins;
— Declaring victory too soon;
— Neglecting to anchor changes firmly in the corporate culture.

These errors can be mitigated and possibly avoided. The key lies in understanding why organizations resist needed change, the multi-step process to achieve it, and how leadership is critical to drive the process in a socially healthy way.

2.8. TRANSITION MANAGEMENT

In any change there is always a future state — a condition one wishes to achieve; a present state — the current condition in relation to the desired state; and a transition state — getting from the present to the desired state, the period during which the actual changes take place. The activities of the transition state may or may not look like those in the future or present states. In managing the overall organizational change process, it is important to determine the major activities for the transition period and to determine the structures and management mechanisms necessary to accomplish those activities. This section will focus on three aspects of transition management: activity planning, where to intervene first and management structures.

2.8.1. Activity planning

An activity plan specifies the critical activities and events of the transition period: when the first moves will take place, when meetings will be held to clarify roles, what information will be communicated to whom and when, and when the new structures will start to operate. The activity plan is the road map for the change effort, so it is important that it is realistic, effective and clear. The following are the characteristics of an effective activity plan:

— *Relevance.* Activities are clearly linked to the change goals and priorities.
— *Specificity.* Activities are clearly identified rather than broadly generalized.
— *Integration.* The parts are closely connected.
— *Chronology.* There is a logical sequence of events.
— *Adaptability.* There are contingency plans for adjusting to unexpected events.

Two considerations in devising an activity plan deserve specific mention: determining where to focus initial attention and selecting the specific change method.

2.8.2. Where to intervene first

Having determined what needs to be changed, the next step is to decide where to concentrate the initial efforts. Any of the following subsystems of an organization can be considered as a starting point for a change effort:

— *Top management.*
— *Management ready systems.* Those groups within the organization known to be ready for the change.
— *'Hurting' systems.* A special class of ready systems in which current conditions have created acute problems.

— *New teams or systems.* Units without a history and those activities that require a departure from old ways of operating.
— *Key subsystems.* Subsystems that will be required to assist in the implementation of later interventions.
— *Temporary project systems.* Ad hoc systems whose existence and life span are defined by the change plan.

There is no best starting point for all change efforts and each change has to be treated individually. The above choices for possible starting points allow questions to be asked systematically and hopefully result in better judgements and better choices.

2.8.3. Intervention methods

In addition to deciding where to start, another issue for analysis involves finding a way to move the change process forward. In deciding upon an initial intervention, one must identify the most promising early activities and carefully think through their consequences. Some of these activities might be:

— An across the board intervention involving all employees;
— A pilot project linked to the larger organizational system;
— Experiments, which differ from pilot projects in that they may or may not be repeated to test different types of interventions;
— An organization wide confrontation meeting to examine the current situation;
— Educational interventions to introduce new knowledge or skills;
— Creating temporary management structures.

One general point to remember is that it is most difficult for a stable organization to change itself, that is, for the regular structures of the organization to be used for managing the change. It is often necessary to create temporary systems to accomplish the change. Change efforts require new ways of approaching problems, as existing mechanisms may be inappropriate or ineffective in such situations.

The choice of the most suitable method for managing a particular change should be decided after a thorough analysis, with an option to adjust it later based on the results obtained or the exigencies of the situations.

The intervention method, where to intervene first and the activity planning are all interconnected and it is difficult to separate one from the other in temporal terms. Therefore, their analysis should happen simultaneously.

2.8.4. Transition management structures

If the transition state is very different from either the pre-change or the post-change condition, a separate management structure compatible with the tasks and organization of resources within this unique state will be needed. The most appropriate management system and structure for the transition state is one that creates the least tension with the ongoing system and the most opportunity to facilitate and develop the new system.

A successful transition manager usually has the following attributes:

— The authority to mobilize the resources necessary to keep the change effort progressing. In a change effort there can be competition for resources with others who have ongoing work to do.
— The respect of the existing senior management team and employees.
— Effective interpersonal skills, as a large part of leadership at times of change requires persuasion rather than force or formal power.

Whatever the choice, making and communicating explicitly the decisions about the transition management structure are important for effective transition. The work to be done during the transition period of a change effort must be viewed on its own unique terms. It needs to be systematically addressed through action plans, carefully monitored through a control system and thoughtfully managed.

2.9. INTEGRATED PROGRAMME FOR ORGANIZATIONAL CHANGE

This section outlines an integrated approach for managing organizational change. The approach is to pursue five different tracks that are scheduled to sequentially follow each other in a particular order. The tracks can overlap; a track does not have to be completed before the next track is initiated. The five tracks are:

(i) Culture track;
(ii) Management skills track;
(iii) Team building track;
(iv) Strategy structure track;
(v) Reward system track.

Figure 4 shows how the tracks might be scheduled against an illustrative timeline.

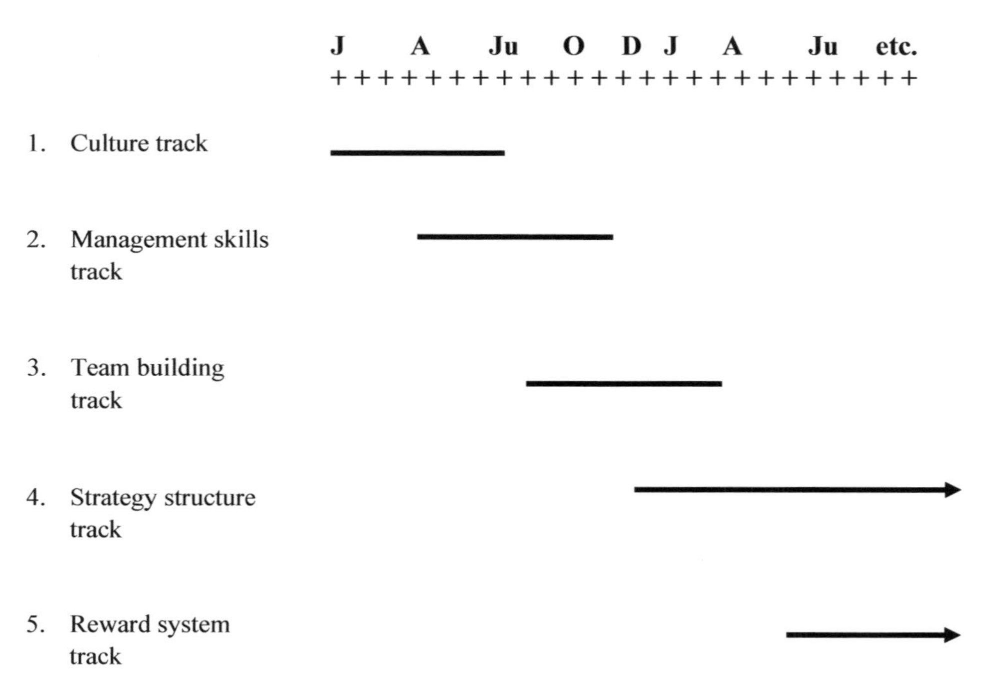

FIG. 4. Scheduling the five tracks.

2.9.1. Objectives of each track

Scheduling the five tracks also requires numerous choices regarding who will be involved at each stage. Typically, the culture track includes every work group in the organization. Every employee's involvement is the only way to change something as ingrained as organizational culture. Every work group should be subdivided into peer groups for each workshop in the culture track. Superiors should be separated from their subordinates in the subgroups to provide the best opportunity for a candid discussion.

Scheduling the management skills track usually involves all the managers in the organization — from first line supervisors through to senior managers, including the senior executives. Just as in the culture track, all group discussions take place in peer groups to foster open communication.

Scheduling the team building track brings the superiors back together with their subordinates in their formal organization groups. This is the only way to ensure that the new knowledge gained from the workshop sessions can be transferred directly to the job. If, however, the superiors and subordinates are brought together too early, before the new culture and skills have been internalized, people will tend to revert to the old ways. It does take some time — in a relatively non-threatening environment — for people to learn new behaviour and skills.

Scheduling the last two tracks, the strategy–structure track and the reward system track, generally involves the formation of two separate task groups. One task group addresses the strategy–structure problem and the other addresses problems with the reward system. The persons chosen for these special task groups need to not only represent all levels and areas in the organization, but also to have demonstrated positive contributions in previous tracks of the programme. Following their deliberations, these two task groups will present their recommendations to top management for improving the organization's strategy–structure and reward systems. Subsequently, these groups can play a key role in helping to implement the recommended changes.

If a programme of planned change is scheduled over too long a period of time, employees will lose interest and become disillusioned because the promised benefits do not materialize. If the programme is scheduled for too short a time period, however, it will be impossible to lay the foundations for the successful completion of each track, and employees may face difficulties or failure as the programme proceeds.

It may be necessary to adjust the schedule as it is being implemented, so there needs to be a degree of flexibility. How long will the process of implementation take? One can expect the first cycle of implementing all five tracks to take anywhere from one to three years. A period of less than one year may be adequate for a small organization.

While the five stages of planned change are complex, so are many of the organizational problems that the change is aimed at solving. The programme must have the support of top management and its implementation must be integrated and flexible. Any attempt to quick fix a programme for planned change is unlikely to succeed.

3. IDENTIFYING AND INTRODUCING THE NEED FOR CHANGE

3.1. POTENTIAL SAFETY PROBLEMS CAUSED BY ORGANIZATIONAL CHANGE

Section 2.5.3 described weaknesses that may be present in an organization's safety culture before a planned organizational change effort was launched. This section discusses safety weaknesses that could be introduced by the organizational change and that, prior to the change, were absent. Figure 5 illustrates the systemic origins of latent failures in an organizational system that can result in unsafe acts that could eventually cause accidents.

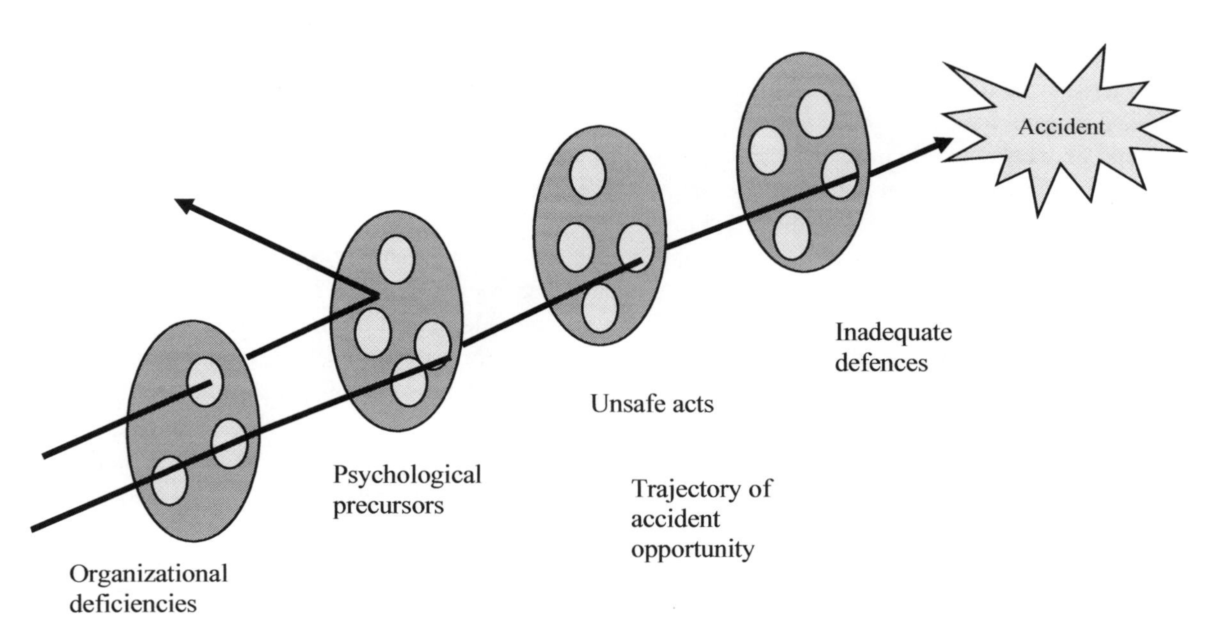

FIG. 5. Systemic causes of accidents: Reason's 'Swiss cheese' model.

James Reason proposed a theory of accident causation that identified a number of elements in the chain leading to an accident. The elements involved latent organizational failures and unsafe acts. The latent failures may be fallible decisions made by organizational decision makers, line management deficiencies or psychological precursors of unsafe acts. Psychological precursors are potential sources of a wide range of unsafe acts. Whether certain precursors will lead to an accident depends on the complex interaction between the task to be performed, the particular dynamic environment and the hazards present at the time. Examples of precursors are: inattention, undue haste, stress, high workload and overzealousness. Unsafe acts are the acts performed by employees that are the causes of accidents and incidents. Some of these are: attention failures, such as omissions, incorrect ordering and mistiming; memory failures, such as omitting planned steps or forgetting; and judgement failures, such as misdiagnosis, misperception of hazards or cutting corners. The unsafe act is the last element in the chain that starts with latent failures.

Systems have defences built up during the life of the organization as it learns from past incidents or risk assessments. But frequently there can be a hole in the defence known as a limited window of accident opportunity. A complex combination of latent and active failures is necessary for the trajectory of accident opportunity to find a hole in each and every layer of defence. The chances of an unsafe act or broken down defence resulting in an accident are small but, unfortunately, very unpredictable.

3.2. PRACTICES TO COUNTER SAFETY PROBLEMS DURING ORGANIZATIONAL CHANGE

The previous section described some of the potential safety problems that can be encountered during organizational change. This section describes some of the practices that an organization can adopt to promote workplace safety during change. The practices are described below:

— *Providing transformational leadership.* Transformational leadership provides an appropriate leadership model for demonstrating commitment to safety and in turn enhancing workplace safety. Transformational leaders are able to act as role models. They are highly respected because they do what is right and not necessarily what is easy or personally beneficial. They are able to inspire their employees to work for the collective good of the organization. With respect to workplace safety, transformational leaders are able to convey to people the value that they place on safety and are capable of encouraging employees to look at safety problems from different perspectives.

— *Measuring variables critical to success.* Optimal measures will provide information that is useful for interventions. Reactive measures of numbers of incidents or safety infractions can neglect the opportunity to learn and enhance safety in the organization. With respect to safety, focusing on current safety conditions and employee attitudes and behaviours that predict subsequent safety performance would provide more relevant information for the prevention of future safety incidents than would focusing on past safety incidents. This is not to say that the number of safety incidents should not be considered but rather that considering process oriented measures would provide an organization with much richer information regarding safety. For instance, it would be worthwhile to measure the behaviour of leaders and employees' commitment to the organization, job satisfaction and trust in management, and the extent to which workers take the initiative with respect to safety and participate in safety matters.

— *Providing training.* Training is a crucial aspect of any human resource system. Employees who have undergone safety training experience fewer accidents than their untrained counterparts. Safety training is especially important in work that is inherently risky, given the high cost of an error and the inability to learn by trial and error. However, the potential benefits go beyond the training itself. It is important not only that employees are well trained, but also that they see that management is committed to safety training. Training also has the added benefit of increasing organizational commitment.

— *Sharing information.* Information is one of the most important organizational resources, and providing employees with information allows them to understand the purpose and goals of a change effort. Other than simply giving employees the information they need to work safely, information sharing may also impact safety by ensuring that employees feel that they are an important part of the organization, and this has positive consequences. Information sharing is also important in a safety context where it would be costly, both financially and personally, to learn from mistakes. Information sharing is critical to learning and to incident prevention.

— *Creating self-managing teams.* Teamwork and the decentralization of decision making can benefit employee performance, including safety performance. Employees working in well functioning teams tend to feel more accountable for safety in general, and for each other's safety in particular. Teams promote the sharing of ideas and this promotion should benefit safety.

— *Ensuring that work is fulfilling.* Work that is fulfilling will ensure that employees are focused, attentive and emotionally engaged. It is important that the workload is appropriate (i.e. work that is neither overly taxing nor boring). Both work overload and underload can adversely affect safety. The provision of greater autonomy is another factor that has been found to benefit workplace safety. Job autonomy increases employee commitment to the organization and safety compliance. Role clarity is also important to workplace safety.

— *Reducing status distinction.* Distinctions in status that convey the message that some members of the organization are more important to its functioning than are others create unwanted barriers between employees and breed resentment and harm motivation. Employees from all levels should feel able to contribute their knowledge and energy to benefit the organization. This is why feedback channels are very important during organizational change. Where status distinctions are high it will be difficult for management to appreciate the problems encountered by frontline employees, and it makes it less likely that employees perceive the extent to which management is concerned with the well-being of employees.

— *Providing employee job security.* Employment security encourages a long term outlook within organizations, promoting trust and organizational commitment. Employment security benefits workplace safety by retaining an experienced and trained workforce. Employment security increases trust in management who do not consider employees as dispensable.

The above eight practices should help maintain the resilience of safety during an organizational change. The maintenance of a positive safety climate during the change process is also important as this reflect employees' perceptions of the organization's commitment to safety.

3.3. SCREENING CHECKLIST FOR MANAGING ORGANIZATIONAL CHANGE

The following checklist can help in identifying issues that require further consideration when planning an organizational change.

(A) Does the change involve a position with potential safety impacts?	Yes	No
Reactor operator		
Line supervisor or manager		
Maintenance personnel		
• Radiation protection supervisor		
• Radiation protection staff		
• Licence manager		
• Safety committees		
Technical support staff for the reactor		
Safety personnel		
Environmental safety personnel		
Emergency response personnel		
(B) With respect to the above positions, does the change involve:		
Reduction in the number of positions		
Reduction in the number of persons in those positions		
Significant increase in duties		
Significant change in responsibilities		

If the answer is yes to any position in (A) above, and/or the change is of a type in (B), then there may be potentially significant safety impacts and a more detailed assessment is required. Guidance on a more comprehensive assessment is given in Section 6.

3.4. ANTICIPATING ORGANIZATIONAL CHANGE

The challenge for organizations is to recognize change as a continuing opportunity, rather than a threat. Changes ignored or not anticipated can have a devastating effect. The key to managing change is not reacting to change too late. The survival and success of an organization comes from anticipating, adapting to and generating fresh ideas that exploit the changing conditions. More than ever, an insight into tomorrow is the difference between success and failure.

3.5. SIX TYPES OF 'EVOLUTIONS' TO CONSIDER

There are six distinct but linked 'evolutions' that influence the nature of business, the structure of organizations, and the behaviour and attitudes of managers and workers alike. Figure 6 represents these six evolutions as linked circles, highlighting the interaction between each of them. Understanding the nature and effects of these evolutions helps organizations anticipate and make decisions on the changes necessary for future success.

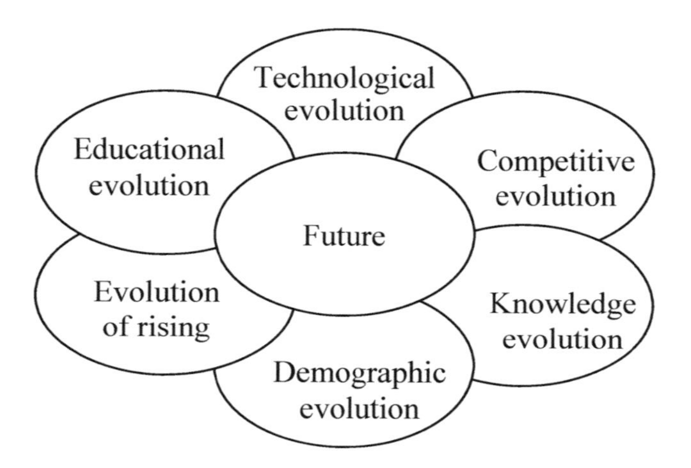

FIG. 6. The six evolutions.

These parallel evolutions do not exist as distinct patterns of change that stand alone. They all interact with, and to some extent stimulate, each other. The most fruitful area for consideration when anticipating the direction these evolutions are moving is to identify possible convergences. The convergence of evolutions is the real change agent. Anticipating change depends on the convergence of thought patterns and opinions. Much of organizational success depends upon the convergence of theory and practice.

Organizations must take ownership of the need for change, understand the need for change, and then implement the change that is relevant to their present and future.

4. HARNESSING SUPPORT FOR CHANGE

4.1. ORGANIZATIONAL CULTURE

The most difficult aspect to manage in any organizational change effort is arguably that of cultural change. As culture seems so difficult to manage, many organizations just ignore it and focus instead on the more tangible aspects of change, such as operations, equipment, systems and procedures. Achieving and sustaining the goals of organizational change efforts requires the culture of the organization to be considered. Effective implementation requires that all changes be clearly connected to an organization's culture. Making this connection not only enables effective implementation, but also embeds change into the daily life of an organization. As a result, change is sustained and desired results achieved. There are various 'lenses' through which we can view the organization, and the choice of lens will affect the way that we approach organizational change and the issues that we perceive as deserving priority action (Table 2).

TABLE 2. THE LENSES THROUGH WHICH AN ORGANIZATION IS VIEWED

Lens	Examples of what is noticed or discussed
Structural	Goals, objectives, roles, responsibilities, performance, policies and procedures, efficiency, hierarchy and control
Human resource	Participation, feelings, fulfilment, communication, needs of people, relationships, motivation, enrichment and commitment
Political	Power, conflict, competition, authority, experts, coalitions, allocation of resources, bargaining and decision making
Symbolic	Rituals, culture, values, stories, different perspectives, language, expressions, myths, commitment and metaphors

The likelihood is that all of the lenses will be used at some stage during the change process. It may help avoid conflict if those involved in the change process are aware of which particular lens they are using. It is important to realize that no lens, or perspective, is better than the others. Usually, the more lenses through which you can view the organization, the better it will be for the chance of success in an organizational change project.

4.1.1. Components of an organizational culture

Culture can be viewed at different levels: the artefact or observable level; attitudinal or behavioural level; and finally the deepest level of beliefs. From a practical viewpoint it would be very difficult to address all levels simultaneously when embarking on an organizational change. Initially, it is possible to consider what changes can be made at the artefact level to promote the desired organizational change.

Artefacts of culture can be isolated, but no one artefact fully characterizes an organizational culture. Table 3 illustrates ten artefacts that together establish an operational description of organizational culture. By identifying discernible artefacts of organizational culture, we can determine the tangible elements that may be managed to help implement and sustain change. However, just as no single component of the ten identified in the table defines culture, it will not suffice to involve only one in support of a desired change.

The ten components listed in Table 3 together can be applied to the change process as a cultural screen, as shown in Fig. 7. The best way to apply the screening process is first to choose a proposed change, and then identify any aspect of the change that is associated with any of the ten cultural components shown in Fig. 7. For example, consider an organizational change that involves the merger of two departments. The screen can be used to systematically consider the potential influence of any of the ten cultural components on the proposed merger.

TABLE 3. CULTURAL ARTEFACTS

Rules and policies	Leadership and management behaviour
Goals and measurement	Rewards and recognition
Customs and norms	Communications
Training	Physical environment
Ceremonies and events	Organizational structure

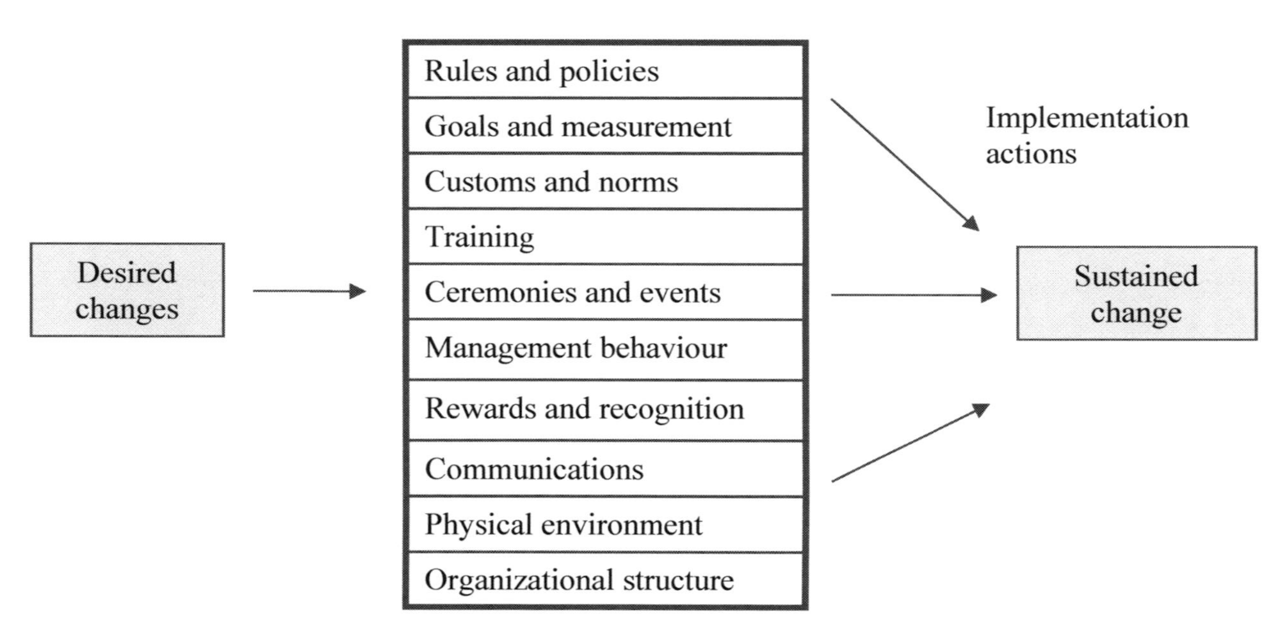

FIG. 7. Applying cultural artefacts to the change process.

For example, are any of the organization's rules and policies, goals and measurement, training, management behaviour, communications, physical environment, or other parts of the organizational structure associated with the merger? In this example, the answer will almost certainly be affirmative. Old rules and policies may need to be replaced by new ones that apply to the merger. Management and employees may need new goals and measurements to support the merger. Putting a desired cultural change through this screening process enables organizations to identify the cultural components that will facilitate implementing and sustaining desired change.

After the screening process, several implementation actions should be developed to manage each cultural component as part of the change process.

Table 4 illustrates examples of implementation actions for each of the ten cultural components.

4.1.2. Implementing the actions

Once appropriate implementation actions are developed for each cultural artefact needed to reinforce the desired change, an implementation action plan should be prepared. The action plan should focus on successfully leveraging each cultural artefact to implement and sustain the desired changes. The plan should include the people involved, the timing of implementation, methods of communication, interested parties and the resources needed. After implementation has begun, the impact of making the changes should be measured. There are two main reasons for doing this. First of all, an organization must be able to determine when goals for change have been

TABLE 4. IMPLEMENTATION ACTIONS

Cultural components	Examples of implementation actions
Rules and policies.	— Review rules and policies and identify those that will hinder performance of new methods and procedures. — Create new rules and policies that reinforce the desired ways of operating. — Develop and document new operating procedures.
Goals and measurement.	— Develop goals and measurement that reinforce the desired change. — Make goals specific to operations by establishing goals and measures for employees who conduct the process being changed.
Customs and norms.	— Identify old customs and norms that reinforce the old ways of doing things and replace with new customs and norms that support the change.
Training.	— Introduce training that supports the change and eliminate old training that is incompatible with the new way of operating. — Deliver training just in time so that people can apply it immediately. — Develop experiential training that provides hands-on experience with new procedures.
Ceremonies and events.	— Establish ceremonies and events to reinforce the new ways (e.g. team award ceremonies for teams that achieve goals).
Management behaviour.	— Identify the desired management behaviour. — Develop goals and measurements that reinforce the desired management behaviour. — Provide training to management that focuses on the new behaviour. — Recognize and reward managers who change to the desired behaviour. — Coach managers who do not change behaviour and convince them that they have to change behaviour.
Rewards and recognition.	— Eliminate the old rewards that reinforce the old ways and replace them with rewards that support the change. — Make rewards specific to change goals that have been set.
Communications.	— Eliminate communications that reinforce the old way of operating and replace with communications that support the change. — Deliver communications in new ways to show commitment to the changes (use multiple channels to deliver consistent messages) before, during and after the change is effected. — Make communications two way to obtain feedback.
Physical environment.	— Establish a physical environment that reinforces the change (collocate managers and employees who will need to work together to make change successful; use telecommunications to connect people who need to interact despite being geographically apart).
Organizational structure.	— Establish an organizational structure that will reinforce operational changes (e.g. eliminate management layers; centralize or decentralize work as needed; combine overlapping departments).

achieved. Secondly, measurement provides a way to track progress and to determine whether the organization is on the road to achieving its change goals.

A complete picture of the effectiveness of changes made can be obtained through a combination of quantitative and qualitative measures. Measures of effectiveness that go beyond financial performance include behavioural observation of employees. In addition to behavioural observation, management and employee feedback can help determine the impact of changes made. Management and employees are often the best sources to discover what is working well and what is not. As with any measurement, steps need to be taken to assign accountability for gathering information, setting the frequency of collection, analysing and reporting information.

4.1.3. Reinforcing change by managing cultural components

The common means for accomplishing cultural change seems to be a combination of the following:

- — Leadership that communicates cultural values in talk and action;
- — New recruitment and selection procedures so that people expected to be supportive of a desired change will be hired;
- — New forms of training to reinforce the desired values and beliefs;
- — Performance appraisal systems in which the culturally correct ways of behaving are rewarded and encouraged;
- — Promotion of persons expressing and symbolizing the desired culture;
- — The use of organizational symbols to signal what is important.

To ensure that change goals are achieved once implementation and measurement have both started, the cultural components should continue to be managed. Continued management of the components during a change effort is essential to reinforce and embed change into the daily operations of the organization. If this continued management does not happen after changes begin to be made, implementation efforts slip. The progress of change will not be sustained. The ongoing management of the cultural aspects should be built into the implementation action plan. The plan should identify who will be responsible for continuing the management of the cultural components. The plan should also identify the time frame for continuing management, and the resources needed, e.g. training facilities to continue education, budgets to continue communications, and individuals to update regularly the goals and measurements and rules and policies of the changes made.

4.1.4. The challenge of changing organizational culture

Planned organizational culture change is generally recognized as a difficult project. Cultures are in flux, but frequently change slowly, and ongoing changes are sometimes not what were planned. Intended culture changes call for creativity, insight, coherence and considerable persistence. Cultural change calls for receptiveness among employees to new ideas, values and meanings. Without such openness — which may be facilitated by cultural changes in society, or growing awareness of fundamental problems in the organization — radical, intentional cultural change is very difficult.

It may be helpful to assess an organization's readiness for change as part of the preparation process. The questionnaire in Annex I can be used to evaluate the degree of readiness.

4.2. REGULATORY ROLE IN ORGANIZATIONAL CHANGE

4.2.1. Regulatory role in a changing environment

Nuclear safety regulators are confronted with two objectives that must be reconciled:

- — Increasing interaction with the public and disseminating information on licensee performance and regulatory actions.
- — Ensuring that the licensed nuclear facility (licensee) takes primary responsibility for safety.

The first objective emphasizes the flow of information and dialogue with the public on regulatory issues. This tends to increase the importance of providing easily understandable measures of satisfactory safety performance and immediate explanations of any licensee anomalies. The second objective emphasizes the importance of safety goals and the responsibility of licensees in meeting them. Compliance, while still fundamental, varies in short term importance according to the safety significance of the regulation or standard.

Nuclear facility owners often seek a more flexible commercial framework for facility operation. This may involve corporate alliances, including foreign ownership. Also, the nuclear power industry and the societal context within which it is regulated are dynamic. Therefore, regulators should continually assess their approaches to regulation to best achieve their regulatory mandate. This includes adjusting the boundaries of activity between the licensee and the regulator; that is, working out the practical approaches that allow the licensee to achieve and maintain safety while allowing the regulatory body to assure itself and the public that safety is not compromised.

4.2.2. Thresholds for regulatory involvement

The challenge for governments in the regulatory area is to maintain high standards of safety while ensuring that regulation is effective and focused on real risks. Regulators also have to find ways of assessing and encouraging the role of senior management in fostering the safety culture in facilities, and in maintaining a focus on safety as a high priority over time.

The main issue for licensees is how to make sure that their safety management is both effective and efficient. There is increasing demand by citizens for even higher levels of safety, together with the demands for the benefits that can arise from a deregulated environment. The environment for managing and implementing change is therefore challenging, and requires a thoughtful approach to organizational change that can only be achieved with awareness of public sensitivities.

The regulator plays a significant role in fostering or impeding the licensee's learning processes. Governments should also consider their role in maintaining or strengthening public confidence in safety authorities. Nuclear regulators should adapt both to the call for greater regulatory effectiveness and to the new conditions of changes such as electricity market competition. Regulators should adapt to changing licensee technical operation and commercial arrangements within the nuclear industry. For example, electricity market competition will lead to increasing demands for regulatory authorities to adopt smart regulatory strategies that incorporate a stricter application of the principle that regulatory authorities should limit the scope of their actions to technical and human organization issues of the licensee that can have a direct and clear impact on the public and workers. Licensees will wish to reinforce their ability, without undue regulatory intervention, to make decisions on how best to protect investment, while citizens will wish to ensure that there is adequate regulatory oversight.

4.2.3. Regulatory tools

There are general principles of good regulation that nuclear safety regulators should use for their activities, including regulation of changes:

— *Accountability.* Regulators must be able to justify decisions and be subject to public scrutiny.
— *Transparency.* Regulators should be open and fair, and keep regulations simple and user friendly.
— *Predictability.* Regulators should set clear safety requirements.
— *Consistency.* Rules must be joined up and implemented fairly.
— *Proportionality.* Regulators should only intervene when necessary. Remedies should be appropriate to the risk posed.
— *Targeting.* Regulations should be focused on the problem and minimize side effects.

In the assessment of the management of change regulatory authorities should:

— Develop criteria and tools for regulatory judgement;
— Clarify regulatory expectations and objectives concerning management of change;
— Make explicit regulatory interest in the management of change;
— Adopt a pragmatic approach and not impede legitimate business practices;
— Encourage licensee self-assessment;
— Promote assessments by the IAEA, the World Association of Nuclear Operators (WANO) and the Institute of Nuclear Power Operations (INPO);
— Focus on the steps taken to maintain knowledge, competence and the ability to act as an intelligent customer;
— Ensure that proper oversight is maintained of the outcome of the change process, and of the monitoring and review processes applied by the licensee;
— Maintain dialogue with licensees to ensure timely and effective scrutiny without imposing an unnecessary burden;

— Consider how to develop and resource expertise within the management of change processes;
— Maintain dialogue with other regulators to share experience (recognizing appropriate national differences).

4.2.4. Assessment of the competence and capability of licensees

Regulators should ensure that organizational changes have been evaluated and classified by the licensee. They should then make a judgement on whether a proposed organizational change will threaten a licensee's competence and capability. Licensees should at all times retain a sufficient level of competence to understand and control the risks of the nuclear facility. The outsourcing of activities is an example of such a threat, and a licensee will need to retain the capability to oversee and control the outcome. Additionally, such outsourcing can erode corporate memory and knowledge; therefore, the regulator may require the licensees to address this issue in a transparent manner.

An effect of deregulation is that licensees seek to become more efficient by reducing their cost base. Invariably this is achieved by downsizing to reduce staffing levels. There are some areas of concern for the regulatory body:

— How the licensee maintains the capability to act as an 'informed customer/buyer';
— How the licensee manages the knowledge transfer of employees leaving the facility;
— How and what controls should be applied to the use of contractors in key roles and activities and the extent of their use.

Regulatory bodies may require licensees to substantiate their organizational structure and their staff's competence and capabilities. Such baseline substantiation should be used in the management of the change process in the same way that the design basis is used in the evaluation of engineering change. Regulators may also expect a periodic review of this baseline substantiation.

There are examples of appointments to senior levels in nuclear facilities of persons who have little or no background/knowledge of the nuclear industry. While the recruitment of senior personnel with diverse business skills is potentially beneficial to the nuclear industry, it is important that the recruits have safety credibility from the perspective of the industry and regulator. The senior managers of the licensee should remain aware that they have ultimate responsibility for safety and that they should be convinced that safety considerations have been given priority commensurate with their significance during any process of major change.

It is essential that managers understand the consequences of their decisions and actions on safety. It should be considered desirable for most members of senior management of a nuclear organization to have knowledge of subjects such as the international safety conventions, safety culture, nuclear regulations and licensing, and industry behaviours and practices.

4.2.5. Measuring, monitoring and inspecting by regulators

The regulatory body must take responsibility for ensuring that its own assessment processes are robust and adequate, and can influence the licensee to develop solutions. The extent of involvement of regulatory bodies in the change processes of the licensees under its jurisdiction will vary between countries depending on the type of regulatory regime employed. However, the level of involvement of regulators will depend on the safety significance of proposed changes (and this is what may be difficult to predict for organizational changes).

Regulators need to have their own monitoring activities to oversee safety during change and detect deterioration. In particular, they have to focus on the long term outcomes of changes, and ensure that licensees are not seduced by short term benefits in efficiency to the detriment of safety and the licensees' technology base in the long term.

It is necessary to continue with the development of regulatory approaches to the assessment of safety implications arising from organizational changes.

It is important for regulators not to impede or unnecessarily slow down changes that are beneficial, neutral or minor in their safety significance, and limit their involvement to the impact of change on safety. Systems to monitor and assess the safety impact must not be bureaucratic. Further, it is important that regulators remain open-minded and do not take on responsibilities for decision making on managerial issues which properly rest with the licensee.

Regulators may incorporate into their inspection plans ways of inspecting a licensee's change process to test the effectiveness of the process by evaluating the outcomes of individual changes. Some regulators have been given stronger powers by their government to regulate the change process — for example, to oblige licensees to have management of change arrangements and to prevent changes which, in their opinion, may produce an unacceptable degradation to safety. In this way, licensees are obliged to ensure that there are adequate arrangements to control any change to the organizational structure or resources that may affect safety. The regulator must approve any subsequent change to these arrangements. Adequate documentation dealing with the safety significance of the proposed changes has to be provided to the regulator by the licensee. The regulator, if dissatisfied with the arrangements, has the power to halt the proposed change. Regulators should be prepared to act decisively in such circumstances.

To increase the leverage of its actions, regulatory bodies need to ensure that licensees pay due attention to the 'high level perspectives' of safety management. Regulators need to support implementation of existing knowledge and development and application of new knowledge.

The following list represents several reference areas for scrutiny by regulatory bodies at the corporate level:

- Visible leadership and commitment of senior management;
- Safety role of line management;
- Mutual trust between management and the workforce;
- Strategic business importance of safety;
- Absence of safety versus production conflict;
- Supportive organizational culture;
- Organizational learning;
- Measurement of safety performance.

At the facility level, regulatory scrutiny should cover the following areas:

- Change control process;
- Recognition of weaknesses and programmes to resolve them;
- Standards and their maintenance;
- Event reporting and corrective action process;
- Safety performance indicators;
- Processes used by managers for self-assessment;
- Involvement of all employees;
- Function of corporate and engineering/equipment support and oversight groups.

4.2.6. Mutual understanding through communication

It is important that the licensees and regulators have a clear focus for dialogue. Dialogue helps to ensure a common perception of the importance of specific issues and hence avoid the imposition of perceived unnecessary regulatory burden.

Ideally, the licensee's management of change process should be clear and visible to the regulator. For instance, it may formally acknowledge the regulator's role, and the timing of its interactions with the licensee. In turn, the regulator should express its own views clearly, and there must be an agreed end-point to the change programme. The importance of good dialogue cannot be overestimated.

There is a growing need for a clear focus in dealing with operation and management issues within a licensee's business. Regulatory interest in operation and management is considered quite new to many licensees. Approaches to planning and managing change may appear novel and uncertain to the licensee — who may be more familiar with the approaches to managing technical, as opposed to human and organizational change. It is important that both the licensee and the regulator have competent staff in this area, i.e. management systems/experience and understanding of the human elements of organizational change.

It is recommended that senior regulatory officials meet on a regular basis with senior executives from licensees. Informal meetings need to explore the long term plans of licensees. The regulator needs to use these meetings to gain reassurance about the commitment of the executives of the licensees to safety during organizational change, and establish confidence in the efficacy of the change process being implemented.

Topics for overall discussion can include:

— Significance of safety:
  • Safety as a business priority;
  • Conflict with profitability;
  • Setting targets and standards;
  • Direction, review, correction.
— Business importance of nuclear activities:
  • Significance of nuclear activities to the licensee;
  • Company expectations;
  • Effects of deregulation;
  • Perception of station operation and safety.
— Management of safety as a part of the business:
  • Safety oversight;
  • Accountability;
  • Control procedures and follow-up;
  • Resources, human and financial;
  • Delegation;
  • Reporting and information flow;
  • Current major problems facing the facility.

In addition, regulators should communicate with government departments and other regulators (particularly market regulators) to avoid, where possible, inconsistency of requirements that might be detrimental to safety. Furthermore, regulators should institute regular and open dialogue with the public — their customers — so that issues about proposed changes in nuclear facilities can be discussed.

4.3. COMMUNICATION DURING ORGANIZATIONAL CHANGE

4.3.1. Communication fundamentals

In order to be effective, a communications plan must be guided by several fundamental principles:

— Communication messages should be linked to the strategic purpose of the change initiative. For example, if an initiative's purpose is to reduce costs, messages should explain why cost reduction is necessary, what the goal is, what the benefit of reaching the goal will be, and who will benefit from it. Making the link to the strategic purpose will help establish an understanding of the need to change, keep people motivated and on track during the change process, and establish credibility around the initiative as a positive thing to do for the organization.
— Communications should be realistic and honest. Glossing over possible negatives will create a belief that the messages are not honest. Conversely, honest communication of all aspects, both good and bad, will help people believe the messages. In addition, explaining the goals and limits of a change effort will help prevent people from jumping to conclusions of worst case scenarios.
— Communications must be proactive rather than reactive. They should be planned in advance and begin early in the change process. Proactive communications will help avoid the need for a defensive position during the process. People may resist change if they believe that they do not possess the skills necessary to take on new responsibilities. Communicating that necessary training will be provided will help alleviate concerns and lower the resistance to the proposed change.
— Messages should be repeated consistently through various channels. Repetition and consistency increase the impact of a message. For example, an announcement about change that comes from senior management is often misinterpreted because people receive such messages through their own personal filters that distract

them enough that they do not hear the entire message. Personal filters include thoughts about how one can avoid the change, a focus on personal disagreement with certain aspects of the change, or questions about how the change will affect one's own situation. Multiple consistent messages enable people to internalize the message and more clearly hear all components of it. Channels used can include announcements from management, memos, newsletters, videos and group meetings.

— Two way communications are needed to ensure successful implementation of the changes; that is, feedback mechanisms must be established. Effective feedback mechanisms should focus on four elements. During the early stages when the details of the proposed change are being designed, they should focus on the concerns of stakeholders. After implementation of the change has started, they should focus on what is working well and what needs to be refined. Feedback should also be obtained on whether the goals of the change are being achieved. Feedback also gives the opportunity to look for lessons learned that could be applied to conducting future change initiatives more effectively. Feedback is thus the key element in creating an organization that learns from what it does.

4.3.2. Avoiding common pitfalls

Communication mistakes are often exposed during a period of change as symptoms of poor management of the process. Common pitfalls are:

— Believing that people will only learn about the change from official communications. This is not true, as the 'grapevine' is always active and quick. Even the best kept secrets of business and government are speculated about and reported on in the media. To avoid this pitfall, a communications strategy must be formulated and initiated as early as possible.

— Delegating communications to a communications department. A communications department can be useful in assisting with formulating the message, identifying and setting up delivery channels, and helping to organize communication events. However, management must demonstrate ownership of the messages about the change process. When management tries to delegate ownership, it sends a strong message to the organization that the change is not important enough for management to devote time to it. People interpret this message as a lack of management commitment and, as a result, frequently will not commit to the change themselves.

— Providing a poor description of the change during the communication process. Key aspects of the change are often glossed over, including the fundamental reason for the change, the sequence of events that will be taking place and the people who will be involved. Inadequate description of change often results in implementation breaking down at lower levels and employees questioning management's knowledge of the details.

— A poor communication process or plan. This will result in unclear definition of roles and insufficient follow-up or fine-tuning once implementation begins. Conversely, a carefully designed and executed communications plan serves as an effective vehicle to break down barriers to change and establish buy in.

— Insufficient communications from senior managers. This can result in lack of commitment by middle managers, who may view the changes as temporary or reversible. The commitment of senior management must be shown throughout the communications process in order to send a message to the organization that the changes are an organizational priority. Senior management priorities are communicated not only by what is said but also by what is done. Taking the time to make announcements and conduct question and answer sessions will show the organization the importance of the initiative to senior management.

— Shutting down the communication channels once changes have been developed and are in the process of being implemented. This often happens because by this stage management has usually reconciled whatever its concerns may have been because of its immersion in the development and direction setting of the change. Yet this is often the time when people at the lower levels of the organization need even more communication. These people will have questions that they want answered, concerns they want heard, and suggestions which they think will make the changes work better. All of these issues can be addressed by keeping open the communications channels established earlier in the change process.

4.3.3. Phases of the communication process

The leaders of an organization embarking on change have an important role to play in ensuring that the organizational communication system is ready for the task ahead. The skill of a leader is not that he or she is articulate with words but that he or she knows how to create strong images to motivate people. The purpose of communication shifts as a change initiative takes shape. For example, early in the process the focus of communication is on announcing to the organization what is happening and what the process will entail. But as the process progresses, the focus shifts to the 'big picture', to provide an overview of the changes being developed and identify issues that exist or may arise. Any pilot tests being conducted should be monitored, with lessons learned gathered and included in communications. As the change process advances, communication must become more specific and focus on implications for various parts of the organization and the individuals within it. When the implementation of the changes is progressing, communication shifts to emphasize refinement. Refinement means listening to and acting upon management and employee feedback about how to adjust the changes to obtain the desired outcomes. The four phases of communication for a change effort are:

(1) Building awareness ('This is what is happening');
(2) Giving project status ('This is where we are going');
(3) Rolling out ('This is what it means to you');
(4) Following up ('This is how we will make it work').

Although the communication process evolves as the change initiative progresses, there are some constants throughout all phases. The most important of these is demonstration of senior management support for the change initiatives. Senior management must take time to communicate during all phases of the process. The weight of senior management's input and interaction will have a great effect on the acceptance of change within the organization. Another important constant throughout the process is keeping people aware of how the changes fit into the organization's principles and strategic focus. Doing this will help people see the changes within the context of future organizational success.

Once the phases of the communication process are understood, a specific strategy for communication at each phase must be developed. The strategy should identify stakeholders, objectives in communicating with that stakeholder, the key message, the communication channels to be used, the timing of the communication and the accountability for delivering the communication.

Even when the fundamentals of good communication are adhered to, rumour and hearsay will never be totally eliminated. However, applying the principles to develop and implement a well thought out communications strategy throughout all the stages of a change effort will ultimately help beat the grapevine.

4.3.4. Cross-cultural communication

In the future, particularly in large construction and decommissioning projects or nuclear site management, it may become common for different organizations to be involved. The organizations may represent different national or business cultures, or be from industry sectors that were not previously concerned with nuclear matters and where the business culture is different. Where there are differences, communication will require certain skills beyond what is normally expected of a manager; managers will need cross-cultural skills when working with members of another culture. Managers will need to:

— Recognize cultural differences;
— Understand how different cultural priorities influence communication styles;
— Plan and communicate management messages so that they make sense to members of the other culture;
— Understand the intended meanings of messages sent by members of the other culture;
— Overcome communication problems resulting from language differences or communication styles;
— Identify and address elements of a change which might challenge conventional cultural norms.

There is no doubt that the challenge of organizational change increases significantly when different cultures are involved and successful implementation will take longer. The impact of cultural differences should not be underestimated; such differences are considered again in Section 6.

5. BUILDING ORGANIZATIONAL STRUCTURES AND COMPETENCIES FOR CHANGE

5.1. ROLE OF TEAMS IN ORGANIZATIONAL CHANGE

Much has been written about the value of teamwork and the desirability of developing teams. Sometimes the assumption is made that teams are a panacea for all difficulties of management. Such an assumption arises from treating teams as an end in themselves rather than as a very useful means for getting things done in certain circumstances.

Obviously encouraging teamwork, in the sense of effective cooperation among those involved, is helpful as soon as coordination of any sort is necessary. An essential feature of good teamwork is that members of the group know enough about each other for the contribution of each to be such that it employs the skill and characteristics of each person to the best advantage. Acquiring this knowledge is not easy, particularly in the normal circumstances that apply in a typical organization, where the normal day to day working contacts are confined largely to exchanging information. Where groups take time to review the way that they work together and plan to improve that process, the individuals gain a wider knowledge of each other.

5.1.1. Essential role of participation

Participation is the starting point for creating employee involvement in the process of change. Employees should be encouraged to participate by offering them the opportunity to contribute their ideas, express their concerns, and contribute to the ultimate choice of solution or merely influence the choices made. Employees must understand that they have the opportunity — in fact the duty — to influence decisions.

However highly motivated a manager or a group may be, they cannot change the organization without involving the majority of those colleagues who will be most affected by the desired changes. One way to increase participation is to create working teams involving people from different functions and at different levels. These can be regarded as expert teams or support teams whose role is to devise the solutions to be put into effect for each improvement initiative and to ensure that they are carried through.

The solutions sought are often complex and require the use of cross-functional groups. This first step, once achieved, needs to be built on, first, because the number of persons in the expert or support groups will represent only a small proportion of employees who will eventually find themselves involved in, or affected by, the change, and second, because it is important to avoid at all costs creating two categories of employees in the organization — the select few who will participate in the change process using novel methods, and the rest who will watch from the sidelines and wonder what is going on. This rift would inevitably deepen with time and, however good the expert group's recommendations may be, most employees would be unlikely to identify with them or be motivated to put them into effect.

An approach must be devised which allows all of the employees to participate. One method is to create teams made up of employees who are open to the idea of participation. These teams can be trained in working methods that promote widespread participation in the change process. In this way the involvement of employees can be spread beyond the activities of the expert or support teams into the organization at large.

5.2. LEADERSHIP AND MANAGEMENT ROLES IN CHANGE

Leaders and managers who want to create effective teams for implementing organizational change should coach, foster and instil the following attributes, behaviours and characteristics in their teams:

- Sharing knowledge, experience and expertise;
- Listening to the ideas of others and trying to understand them, analysing all the ideas, opinions and concerns put forward so that the best solution for the organization is chosen;
- Working towards consensus by taking care to choose the best solution for the organization without continually accepting poor compromises to keep everyone happy;
- Remembering that the solution does not really belong to the team, but to the rest of the organization as well, so others need to be called upon to improve and endorse the solutions identified;
- Bringing in experts to solve specific technical problems that are beyond the competence of the team;
- Defining clear roles within the team;
- Ignoring hierarchical barriers within the team;
- Defining the objective, results, timescale and responsibilities for the team.

5.2.1. Critical conditions for achieving a high level of participation

The main requirements are as follows:

- *Guiding without attempting to control.* The change process can generate many ideas, initiatives and activities through the work of teams and from encouraging participation of employees. It is unrealistic to control the activities of every single employee. It is better to rely on the creativity of individuals and their ability to make the most of their skills and expertise, and to direct this intelligence towards the objectives of the change by allowing employees some flexibility to propose the best solution.
- *Handling expectations at the lower levels of the organization.* It is often at the lowest levels of management — supervisors and junior line management — that there is the greatest reluctance to give employees the necessary autonomy. They may view any freedom given to their rank and file and the resulting reduction in their control as a loss of authority and power. The supervisors and junior line managers need to have their roles redefined in the change process to one of team leader and coach.
- *Following up on ideas and initiatives.* Once employees have been invited to participate, it is important to make sure that their ideas are taken into consideration. The dynamic of change would soon be destroyed if employees saw their ideas being filed away and not followed up on.
- *Granting the right to make mistakes.* Participation rates will be higher if employees realize that they have the right to make mistakes. They will feel encouraged to express ideas and try out new ways of working as long as they believe themselves to be reasonably well protected in the event of failure. This helps contribute to the effectiveness of coaching and training, since if employees are engaged in continuous learning, they must apply their newly acquired ideas or expertise without fearing the risk of failure.

5.3. TRAINING AND COACHING

Change demands an upgrading of employees' knowledge and skills in two respects. On the one hand, the nature of their activities and responsibilities may change significantly, making it necessary for them to acquire new expertise in the way they actually fulfil their role or do their job. On the other hand, these same employees will also be on the frontline of the change process, so if they are to contribute to change as effectively as possible, they will need to know how to implement change. The aim of training is to help the workforce rapidly acquire these two types of skill.

Change also requires alterations in the way people behave, and training alone cannot achieve this. It will be necessary to give some key people in the organization (especially managers) one to one support to help them accept change and transform their methods and behaviour in line with the objectives of the organizational change effort. This type of support is known as coaching.

5.3.1. Assessing training and coaching needs

The matrix shown in Fig. 8 can be used to analyse the need by deciding where all the employees involved in change fit in the needs analysis.

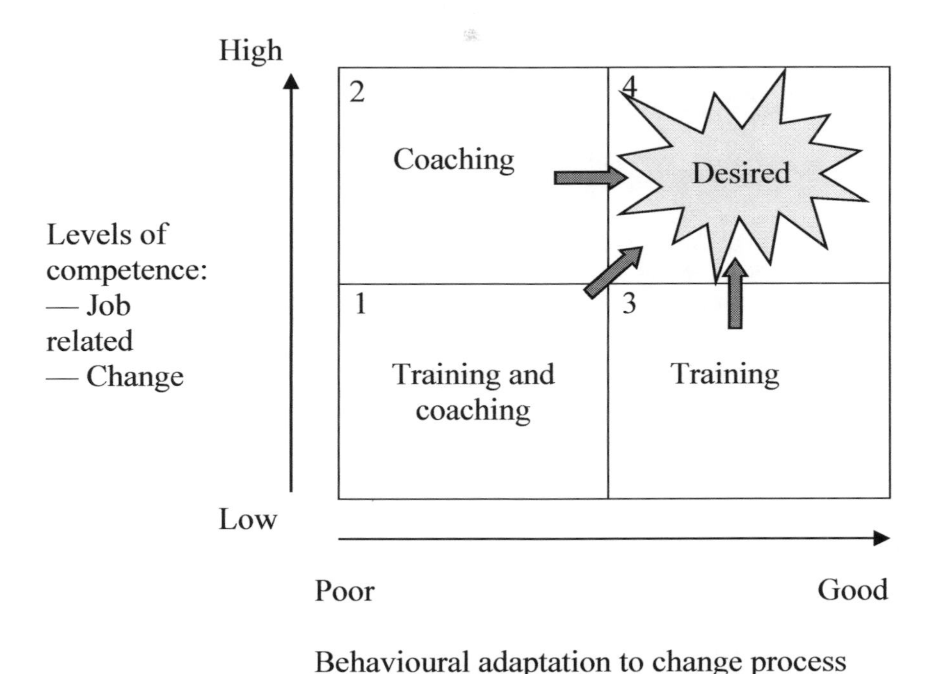

FIG. 8. Matrix for evaluating training and coaching needs.

On the vertical axis of the matrix, the job skills include all forms of expertise related to a role or doing a job. Changes may require developing technical skills, or acquiring greater management skills or broader business skills. The change related skills will allow employees to contribute to the change process as effectively as possible (by understanding the dynamic for change, handling resistance to change and group problem solving). While the skill need will differ according to an employee's role in the organization, almost all of the change skills are directed towards improving the quality and effectiveness of interpersonal relations.

The horizontal axis of the matrix represents the appropriateness of employee behaviour, both to the process of change and to the way the changed organization will work. Behaviour related to change may be a matter of accepting change and being willing to take part in it. Behaviour related to the way the changed organization will be may mean greater readiness to delegate and more self-confidence.

5.3.1.1. Square 1 of the matrix: Inadequate skills/inappropriate behaviour

Employees in this situation do not have the necessary technical skills to take on board their new responsibilities. Shortcomings of this kind are usually easy to identify. In many cases employees also lack the interpersonal skills so vital to the change process, such as the ability to work in a group or to form a project team. To ensure that the change is implemented effectively, some training in communication and cooperation may be required for employees involved in the change process.

The behaviour of employees in this square of the matrix is also inappropriate for the change process or incompatible with the vision of the changed organization. The behavioural development of employees can be speeded up by feedback or coaching. Behaviour, which is at odds with the course of the change process, is often linked to employees' anxiety about not being competent to meet their new responsibilities or to the fear of losing status. Training reassures them and restores their confidence, and thus itself acts as a trigger for behavioural change.

5.3.1.2. Square 2 of the matrix: Adequate skills/inappropriate behaviour

Employees in this situation have the necessary skills for self-development but their behaviour is either opposed to the change process or incompatible with the ultimate objectives of the change. The skill base for change is there, but they may refuse to change or to understand the need for a significant change in their behaviour or attitudes. The aim of coaching will be to help these employees to become aware of their behaviour and attitudes, to analyse them and to change them. Depending on the role of the employee, coaching may take the form of formal one to one support, or it may be less intensive and more informal.

5.3.1.3. Square 3 of the matrix: Inadequate skills/appropriate behaviour

Employees in this situation are motivated by the objectives of the change and enthusiastic about the change process. They want to take part in the change effort and their behaviour is appropriate, but their skills are inadequate. Here the task is to identify the new skills which they need to acquire and to offer them suitable training.

5.3.1.4. Square 4 of the matrix: Adequate skills/appropriate behaviour

These employees have the necessary skills and have adopted the appropriate behaviour. They are already motivated by the changes taking place, but they could be encouraged to give more of themselves. They may be capable of being trained as a trainer or as a coach for their colleagues. These employees can serve as role models to whom their colleagues can look to for an understanding of the ways of working and behaviour expected of them.

5.3.2. Training in change management and level in the organization

The training made available will depend on the level of responsibility of the people and their role in the organization, as shown in Table 5. As shown in the table, group problem solving methods concern everyone involved in the organization, while other aspects of organizational development concern only a small group of decision makers. A training plan should be prepared as part of the change process, as soon as the change objectives have been defined. The plan should cover technical, job related and change related training.

5.3.3. Coaching

In the context of a change programme, the aim of coaching is to give one to one support to the managers or decision makers who, because of their role in the organization or the responsibilities they exercise, play a key role in making change successful. Coaching helps them to reinforce the aspects of their behaviour which promote the smooth running of the change process and the attainment of its goals, and on the other hand to correct or stop those aspects which obstruct or hinder change. The aspects of behaviour (or practices or attitudes) which coaching is concerned with include:

— The capacity to cope with ambiguity and complexity inherent in any change process;
— An enhanced expertise and a stronger command of oneself in the resolution of conflicts;
— An intrinsically confident attitude;
— A stronger inclination to accept the inherent uncertainties involved in change;
— A greater capacity to trust people, delegate to them and make them accountable;
— A greater capacity for listening;
— More efficient time management;
— The ability to overcome the fear of change.

Coaching takes the form of a series of joint sessions, including the coach and the manager being coached. These sessions concentrate on analysing the situations that the manager has encountered, on exchanging views and on sharing experiences. The coach has the role of a catalyst in helping the manager to recognize his or her strengths and weaknesses and then to find solutions appropriate to the particular circumstances. This approach is

TABLE 5. POSITIONS AND THEIR CHANGE MANAGEMENT SKILL REQUIREMENTS

Position	Change management skills required
Organizational leaders.	Develop a vision for the organization; Create and manage a change programme and understand the levers for change and the obstacles. Set up a system for communicating change. Motivate an organization to change. Recognize and handle resistance to change. Steer change. Work in groups.
Departmental and middle managers.	Communicate the organization's vision to employees in their area of responsibility. Manage change according to a programme. Recognize and handle resistance to change. Motivate their employees to change. Run effective meetings. Solve problems in groups. Set up a system for communicating change. Lead group analyses of processes and practices. Create an appropriate structure for the change.
Supervisors.	Recognize and handle resistance to change; Build a team; Run effective meetings; Solve problems in groups; Lead group analysis of the current situation.
Employees.	Solve problems in groups. Run effective meetings. Analyse a process or practice.

based on the principle that people do not learn new attitudes and ways of behaving by being given advice or having ready made solutions imposed on them, but by analysing their own experiences, by drawing lessons from the root causes of their successes or failures, and by looking for their own solutions on the basis of their own analysis. The coach, by helping in an objective way with the analysis, speeds up this learning process. For one to one coaching of managers it is best that a coach should be found from outside the organization (generally an external consultant), because of the independence of mind and objectivity that he or she can bring to bear on the organization's internal issues, the more so in the context of managing change, where power issues and emotional reactions can be particularly heightened.

5.3.4. Development of employees during change

All employees (not just managers) involved in the process of change will benefit from a process designed to help them make progress and develop their potential. Given the financial and human resources that it requires, one to one coaching will be limited to a few key participants in the change effort. Group workshops are a less intensive form of coaching and can speed up the development of all employees by encouraging them and supporting them in analysing their performance and behaviour during the implementation of change. The coach in this case can be another employee of the organization who is competent to act as a facilitator of the workshop. Group workshops can help with the following issues:

— Aptitude for group work;
— Ability to listen and understand the ideas of other persons;
— Ability to motivate others;
— Capacity to grasp the positive implications of events first, and then tackle the potential problems;
— Willingness to accept novelty without rejecting it without consideration;

— Constructive approach to meetings;
— Willingness to help colleagues to succeed;
— Capacity to recognize and reward effort;
— Ability to communicate, explain and persuade.

Coaching and group workshops can be major drivers of behavioural change and development of potential in both the managers and employees responsible for implementing change.

6. IMPACT OF CHANGE ON SAFETY AND SAFETY DOCUMENTATION

6.1. INTRODUCTION

This section provides an essential link between the theory and practice of management of change (MOC). The first part of this report is a theoretical study about making a change to an organization, while the second part takes the form of a practical guide. At this point it is essential to consider the impact of any change on an organization, and to stress the importance of a thorough review of the real implications of the change that has been conceived.

It is demonstrated above that MOC is the process used to review all proposed changes to the management organization and safety documentation before they are implemented, so that their effects on safety vulnerabilities can be adequately determined and addressed.

A change to the organization will have an immediate and obvious impact on personnel and resources, but any review of a change must also consider impacts on the safety documentation; safety documentation can have hidden or implicit requirements on the organization that may be affected by a change. Assumptions made in preparing safety documents may rely on particular organizational or managerial arrangements. It is therefore essential that the review process includes personnel or organizations with an understanding of the safety basis of the plant and its safety case. While changes to safety documentation will be managed using other licensing arrangements, unexpected effects can arise if the organization or management system is updated without proper consideration. In some licensing regimes, safety analyses may be affected; in others, the risk is now specifically identified under 'resources'. Similarly, changes to the internal assessment process, safety boards and other senior level arrangements such as licence managers and safety committees come under the jurisdiction of formal MOC. The organization and changes to it are part of the safety system and process, and need to be considered in the same detail.

6.2. REVIEW

All proposed modifications should be subjected to a review process, in the same way as plant and other physical or operational changes are assessed, and this review must be carried out by knowledgeable persons to evaluate the impact of any proposed changes and ensure continued safe operation. The knowledgeable persons should have expertise in:

— Nuclear safety and the safety system;
— Radiological protection;
— Human resources;
— Human factors and industrial relations;
— Operations.

If the review finds that a proposed change would have an impact on or challenge safety, the applicable safety information should be updated and all employees whose job tasks will be affected by the change should be informed and, if necessary, retrained before work resumes as part of the implementation process. Once the risks have been mitigated and the proposed change is determined to be safe, the change should be implemented so that the impact is controlled and there is no adverse effect resulting from the change process itself.

6.3. EXTENT OF CHANGE AND ITS EFFECTS

Management of change is usually interpreted as relating to permanent changes, but temporary changes have contributed to a number of catastrophic events over the years. Also, organizational changes necessarily result in temporary arrangements while the changes are being enacted and before the change is fully embedded. Temporary changes (e.g. abnormal situations, deviations from standard operating conditions, variations in resource levels, untrained or partially trained personnel filling in during absences) should be managed as if they were permanent changes. Written documentation of normal process parameters, use of standard operating procedures, management sign-offs and clear communications are all necessary.

The potential impact of the change may or may not be immediately obvious. The list of possible effects is extensive and it is not the intention to cover them all here, but there are general principles that need to be considered. Changes that appear to be relatively trivial may in practice result in effects that seriously impair the effectiveness of the organization to maintain safety. Examples include the following:

- Changes in responsibilities and reporting lines, including potential clashes of interest and loss of direct, independent safety reporting (at all levels in the business); disruption of established and secure communications paths; changes of scope, arrangements or independence of oversight functions or safety committees; lines of responsibility lost or changed in unexpected ways.
- Loss of suitable and appropriate authority.
- Potential loss of knowledge or expertise from a given part of the organization or function and consequent loss of focus on aspects of the safety case or operational parameters, or transfer of this knowledge outside the organization.
- Changes to departmental organization, rosters and shift rotations may leave key roles unoccupied or lead to 'singleton expertise' (i.e. single person with a particular set of competences), leading to a dilution of expertise and, in some cases, complete loss by technical experts of their competence status because of lack of continuing work in the particular area of work, excessive working hours and fatigue.
- At a high level, change of ownership may result in a shift of control for funding of safety sensitive activities and the line of control taken away from the competent authority.
- Cross-cultural differences between different parts of a given organization, other related businesses, different business and industry sectors, and national or international differences or sensitivities.
- Changes which raise requirements of nuclear safety that challenge established cultural norms and expectations.

Any of these can, in addition to the direct and obvious safety risks, result in demotivation and the risk to morale, which can cause loss of focus of personnel and potentially their transfer out of the department or business. If this is widespread it can result in a significant unplanned change and threat to safety, particularly if succession arrangements are not robust.

Particular key functions are especially sensitive to change, and it is necessary to identify these for special attention. These normally include oversight and internal regulatory functions, safety boards and committees, and changes that affect reporting lines for safety related matters.

6.4. UNEXPECTED CONSEQUENCES

Once the impact of the change has been established, the potential consequences need to be considered. It is necessary to be aware of the risk of unexpected consequences. The consequence of a change that has not been properly assessed can be loss of independent challenge, loss of expertise and 'informed customer'[1] oversight,

[1] The 'informed customer' or 'intelligent customer' is the capability to assess technical or other work but not necessarily the capability to carry out the work. It is used to carry out technical review of work by third parties.

and these effects can cascade through an organization, compromising safety. Such an adverse effect may not be immediate, but may remain latent for many years before manifesting themselves — possibly long after the personnel with knowledge of the change have moved on.

6.5. ROBUST CONSIDERATION

It is a prerequisite for there to be procedures and processes in place to manage change in the organization; it is usually a licensing requirement. The challenge in any organization is to make the process sufficiently robust without being unnecessarily difficult or bureaucratic, lest the process comes into disrepute. Consideration of changes may be simplified and made more robust by using a systematic approach, perhaps using a template to facilitate the early stages in the process. This facilitates grading of the change and may help identify risks, both obvious and unexpected. Such a template might include a description of the proposed change and why it is being made, as well as safety, health and environmental considerations, and might identify the possible need for changes in existing procedures, drawings, safety plans and other safety documentation. All approvals required should be obtained before any change is implemented. If the impact of the change is minor and clearly understood, the checklist alone, reviewed by an authorized person and communicated to affected personnel, could be sufficient. But for more complex or significant changes, and certainly any that will have a more significant impact on safety, a proper hazard evaluation procedure with assessment and approvals by appropriate personnel and at relevant levels in the organization may be necessary. It is therefore essential that a properly graded approach to review and approval is used at each stage and level. Planned changes should be noted so that revisions can be made permanent when changes to procedures and other documents are made. Documents describing the changes should be kept accessible, available to all affected personnel and other stakeholders.

Changes need to be continually re-evaluated to ensure that: (a) the expected benefits have actually been achieved; (b) safety has actually been maintained; (c) there are no unanticipated or unexpected threats or challenges; or (d) contingency arrangements are in place for the threats that emerge.

6.6. IMPACT OF THE FINDINGS FROM THE FUKUSHIMA DAIICHI ACCIDENT

6.6.1. Introduction

This publication was initiated and largely written before the accident at the Fukushima Daiichi nuclear power plant in Japan in 2011. Many reports and official briefings are now available, which include lessons learned and other appropriate material, based on preliminary assessment results. It is recommended that users of this publication consider these when developing or implementing MOC activities and evaluate the safety impact or any additional requirements that may apply to their organization's operations using the lessons learned from the Fukushima Daiichi experience.

6.6.2. Background

The accident, which took place on 11 March 2011 at the Fukushima Daiichi nuclear power plant, was caused by a devastating earthquake of magnitude 9, followed by a tsunami of unprecedented severity. Several nuclear power facilities were affected by the severe ground motions and the large tsunami waves, which in turn resulted in an immediate, and necessary methodology for managing rapid change. Based on the information now available, the concept of MOC and its relationship to managing an emergency has been recognized as being even more important to the industry than was previously thought. The change from normal operations to emergency mitigation is an immediate, on the spot change, but the same considerations should be employed: the emergency arrangements are essentially the 'end point' of the change and the transition is the change; the authority, accountability and 'command and control' capability must be immediate and clearly defined. The principles of this publication may be applied equally to emergency planning arrangements for such an event.

6.6.3. Findings and recommendations

Amongst other organizations, both the IAEA[2] and TEPCO (owner/operator of the Fukushima Daiichi NPPs) investigated the event and presented preliminary findings and recommendations for the further improvement of nuclear safety [9–11]. Some of the recommendations were related to the reconsideration of:

- Design requirements, with particular emphasis on defence in depth, low probability beyond design basis accidents, both singly and in combination;
- Severe accident management for single-unit and, more especially, multi-unit sites, including extended loss of ultimate heat sink and essential supplies, hydrogen management, post-accident monitoring and safety of spent fuel storage.

Some recommendations focused on the establishment of a new regulatory body by the government, and the need for this body to ensure organizational competence in responding to an emergency to take account of the impact on safety, based on effective management of the changing situation during an accident. It would also be important to develop professional competence to provide appropriate advice and leadership to the responsible personnel and relevant organizations that are in charge of emergency response, and to promote management capabilities to make the best use of available resources effectively and efficiently, emphasizing the importance of providing relevant information and putting in place the system routinely, and to be prepared to provide timely and appropriate information nationally and internationally in case of an emergency. These recommendations are also fundamental to managing change in nuclear organizations.

7. PRACTICAL STEPS TO PREPARE FOR AND IMPLEMENT ORGANIZATIONAL CHANGE

7.1. INTRODUCTION

Many of the potentially adverse impacts on safety of organizational changes can be avoided if consideration is given to the effects of changes before they take place. Both the final organizational arrangements resulting from the implementation of the change and the transitional arrangements need to be considered. An integrated management system based on IAEA GS-R-3 [1] can impose such a discipline within an organization.

A management system based on the requirements of IAEA GS-R-3 [1] and supporting guidance, such as in IAEA Safety Standards Series No. GS-G-3.1 [2] and No. GS-G-3.5 [3], IAEA-TECDOC-1226 [4] and IAEA-TECDOC-1491 [5], would be such a management system. This publication requires the organization to justify each organizational change and to plan, control, communicate, monitor, track and record the implementation of each change to ensure that safety is not compromised.

This section is divided into parts corresponding to the major process steps in the change process. Section 7.1 provides the introduction and overview of the change process. In Section 7.2, detailed descriptions of the activities in each step are provided. Since management oversight is a critical activity that is necessary throughout the change process, it is contained in the last paragraph within each section.

There are different models to organize thinking and action regarding the human aspects of the change. In what follows, the approach of Kotter [12] has been taken as a reference, but other approaches may be valid and may be suitable for some organizations as presented in Ref. [13]. In any case, the application of the steps should be graded according to the complexity and extent of the change. The order of the steps does not mean that they have to be performed sequentially; some overlap among different steps may be beneficial.

---

[2] The IAEA is preparing a report which will present an overview of the event and will provide recommendations for further improvement of nuclear safety.

7.2. MANAGING ORGANIZATIONAL CHANGES

There are a number of drivers for change described in this report and in IAEA Safety Standards. They can be characterized in terms of whether those drivers arise externally or internally to the utility.

In general, internal changes tend to result from an organization seeking improvement in economics, safety policy and practice, technological processes, competency systems, etc. (sought changes), whereas external changes tend to be proposed or imposed by the government, regulators, civil society, etc., resulting in higher impact and more complexity, i.e. privatization, divestments, mergers, etc.

At the functional level, it is necessary to:

— Describe and assess the type of change (what is to be changed), the extent of change (potential impact of change), the level of change (grading).
— Take into account a deeper view of potential impact on safety.
— Determine the forces initiating or provoking the change on overall organizational strategy, technological or human processes, markets, mergers/acquisitions, downsizing/upgrading, regulatory, political, etc., identifying enablers and risks for the change processes.
— Define the appropriate intervention strategy considering types of changes, conditions of effective and efficient management for the identified change, how to strengthen the safety culture and determine performance indicators and critical success factors.
— Design the implementation process to foresee effective communication practices, higher involvement of related people and institutions, planning details to build on the change.
— Forecast ways and means to assess the short term results and impact in the long term of the change project.

Nuclear organizations should seek to mitigate additional difficulties by relying on the buildup of updated management systems and practices to be able to forecast and cope more effectively with potential issues requiring change. The organization should establish a formal and effective change management process focused especially on nuclear, radiation and industrial safety. Examples of how these are presented in the annexes; a specific high level model is given in Annex VI.

7.3. IDENTIFICATION OF THE PROPOSED CHANGES

The management must periodically assess the objectives of organizational structure and current complement ('organizational baseline') and use the results for improvement of the system. Goals, strategies, plans and objectives need to be developed in an integrated manner so that their collective impact on safety is well understood and managed. This may take the form of a review of the organizational baseline. The results of such an assessment could identify the need for a proposed change, or there could be an external driver such as a regulatory requirement or new market conditions.

Nuclear utilities need to:

— Understand these drivers;
— Continually monitor and analyse information;
— Develop strategies that enable them to manage both the present and future;
— Anticipate the needs and introduce change proactively.

Any change should drive towards improved goals and objectives; performance and the management of nuclear safety must be considered with every change initiative. It is important to implement an effective dialogue process for all affected personnel and to communicate with other stakeholders, as appropriate, so that they can understand the change and its importance in meeting the goals and objectives of the utility.

Each utility should issue a policy for promoting and managing change that links the vision and values of the corporate level to what is expected and why. This policy on change management must:

— Give priority to safety;
— Address all types of changes;
— Introduce the change management process;
— Promote effective communication.

The starting point for the identification of the need for a change is the determination of current status, as well as the adequacy of the size and structure of the organization. This baseline will then provide a reference basis against which future events can be compared and judged in the proposed change.

The need for change can also be identified based on:

— Performance trending;
— Benchmarking;
— Forecasting and expectation;
— Analysis of internal and external drivers.

The identification of the need for change requires clearly defined process steps and a systematic approach in line with the vision, mission, objectives and strategies of the organization. The formalized process should include risks of identified proposed change.

7.3.1. Description of the need for change

The initiator must clearly define the need for any change, using an established document format. At the beginning of the process, the proponent must identify the need for and describe the proposed change, using the organization's procedures; this must include a description of the organizational start point and the intended end point. A formal proposal will be needed; this is covered later in this section, but at this stage in the process the initiator should identify:

— *The title of the change.* Clearly describe the true nature of the change.
— *Starting point.* The present organization and condition.
— *Reason for the change.* Include reference to the safety and business case objectives and how the change is integrated with other change proposals.
— *End point.* The expected final result of the change, competencies after completion of the change and the revised structure, if appropriate.
— Main drivers and reasons for the proposed change.
— High level risks, preconditions or enablers and countermeasures to mitigate risks.
— Expected date of the change.

Examples are given in the annexes. Under some regulatory regimes, some of the changes will be covered by engineering or other conditions of the license to operate. It may also be useful to identify and consider several different alternative ways of carrying out the change and alternative end points that may achieve the required organizational benefit.

The change should only be deemed to be complete after post-implementation assessment recognizes that the change has been successful or the degree of success identified by the review has been accepted by the management team. The change can then be closed.

7.3.2. Definition of the parameters to assess the success of the proposed change

Each change proposal should include relevant measurable parameters and indicators. Evaluate the progress and success of the change using clearly defined parameters, including indicators related to nuclear safety. These must be monitored during implementation and evaluation of the change and should clearly demonstrate that there is no negative impact to nuclear safety. Use both quantitative and qualitative measures, which should address the change process in terms of efficacy (impact), effectiveness (in relation to the results proposed and achieved) and efficiency (level of productivity with the available resources). The value of these parameters should be described in the change proposal and these parameters could be from already monitored parameters or new defined parameters. This may include parameters or indicators that are already used by the business.

The parameters should include:

- Safety indicators, which will run before the start point and after the end point;
- Requirements from other relevant codes and standards adopted;
- Statutory and regulatory requirements;
- Requirements formally agreed with interested parties (stakeholders);
- Communication and personal feedback (surveys, etc.) as applicable.

Measurable parameters, including safety indicators, create the basis for successful implementation and evaluation of the proposed change. Do not use general parameters that do not relate directly to the change (e.g. unit capability factor) or are not clearly defined or measurable.

7.3.3. Identification of risks, enablers and countermeasures

It is important to state which aspects of nuclear safety are at risk clearly in the proposal, building on the high level risks already identified, together with how they will be addressed; also, it is necessary to identify all of the potential safety risks. A list of critical risks which may be relevant is provided below, but the proposal should also describe other potential risks.

Examples of critical risk areas in which proposed changes can influence activities important to safety include:

- Insufficient resources (numbers, distribution, competence, motivation, etc.) to ensure that activities important to safety are implemented;
- Decrease in efficiency and transparency of management of safety related activities;
- Potential degradation of activities important to safety;
- Loss of knowledge and abilities important for ensuring activities important to safety.

Evaluation of the risks is covered in Section 7.4.2.

Enablers — Actions to minimize risks. Any change proposal should ensure that suitable enablers are correctly identified and assessed. Measures — enablers — should be included to mitigate the risks and reduce the probability of nuclear safety being adversely affected. These enablers should be SMART — specific, measureable, achievable, realistic and timely.

Enablers are the specific activities that will be needed to make the change work. Where there are dependencies between enablers, it should be made clear how such dependencies will be managed. It must also be clear from the change proposal which enablers are required at each stage of the implementation process; in particular, it should be clearly stated which enablers are required before implementation; the tasks and timescales are then included in the implementation plan.

Countermeasures — Actions to correct inadequate or unsuccessful change. If the change fails or does not progress adequately, or there is a negative impact on nuclear safety, then suitable countermeasures needed to be implemented. These should be identified in the proposal, and must be realistic and implementable. These may include the temporary suspension of the change, reallocation of responsibilities, provision of additional resources; reversion to the original structure may be suggested, but this may not be practical and will require some additional resource or justification.

7.3.4. Definition of the guiding coalition

A single individual must be nominated as accountable for implementation of the change. However, this person will need the support of a 'guiding coalition'. This is the group of people powerful enough and sufficiently motivated to lead the change, and who have knowledge and experience of nuclear safety. They should be nominated at an early stage to drive and facilitate the change. The group will normally include managers, together with individuals with particular knowledge of the area being changed and expertise in managing change. In a major reorganization, this may also include clerical and logistical support. It may also be useful to enlist the specific sponsorship of an individual senior manager.

The guiding coalition should be familiar with the integrated management system of the organization. If responsibilities and accountabilities for implementation of change are not clearly defined and well understood, or the guiding coalition is not determined, the change could fail or its implementation could be delayed.

7.3.5. Development of a vision and strategy to implement the change

The change leaders must develop a clear vision to engage people with the prospect of a beneficial change, and then prepare a strategy to manage the change. This will help to ensure that the main steps and tools to reach successful implementation will be met in line with the main objectives, mission vision and expectations of management. The strategy should also develop a sense of urgency — the imperative and the importance of the change together with the need for timely completion.

The strategy, which will be highly dependent on the type of change, should describe the means by which the change process takes place:

— *People.* Interventions that attempt to change the behaviour of employees, their attitude or values and can often entail the challenging task of changing the organizational culture.
— *Structure.* Changes affecting the organizational hierarchy, allocation of rewards, degree of formalization, and addition or elimination of positions, departments or divisions.
— *Technology.* New technologies enabling the change or technology supporting the change.
— *Organizational processes.* Consider changing processes used by the organization to carry out its business: administrative processes such as those for decision making and communication, core operational processes for delivering the products (items or services) of the organization, support processes (e.g. information technology).

7.3.6. Preparation of a formal proposal

At this stage, all of the key requirements will have been identified. It is now necessary to compile these into a single document, a proposal that can be presented for assessment and approval. The proposal must summarize the information described in Sections 7.3.1 to 7.3.5. A checklist should be prepared based on these sections. At a minimum, it should include:

— The originator and sponsor of the change;
— Why the change is necessary;
— The start point;
— The end point;
— The outline of the change;
— Identification of the relevant stakeholders and the roles and responsibilities being transferred;
— All of the risks identified and the enablers required to mitigate them;
— Any additional enablers that will facilitate the change;
— Countermeasures to be used in the event of the change not working, and when they should be deployed;
— How progress and completion will be measured;
— An outline implementation plan;
— Identification of the grading of the proposal, with justification;
— Identification of the degree of management scrutiny and (if appropriate) internal independent assessment and external regulatory assessment (which may depend on the grading);
— Who is accountable for satisfactory completion of the change;
— An implementation date and time schedule;
— A date for review of the completed change.

The outline implementation plan should identify the major tasks, steps, objectives and milestones necessary to enable the change to be taken forward and implemented. One of these may be the preparation of a more detailed implementation plan. This forms part of the change proposal and assessed accordingly. The main steps and milestones must be covered in the proposal and transferred to the detailed plan describing responsibilities later.

A basic implementation plan should include at a minimum:

- Main resources needed;
- Relevant stakeholders;
- Risks facing the utility for the proposed change;
- The ability to influence and manage risks;
- Impact on safety culture/management.

All relevant organizational units that could be affected by the change must be involved in reviewing the change proposal.

The following questions should be considered:

- Is there a clear policy on change management, aligned to the vision, goals and objectives of the utility, that gives priority to safety and which is communicated to personnel and relevant stakeholders?
- Is there a systematic, transparent and rigorous change management process applied with main milestones to all types of change during its implementation?
- Are enablers and countermeasures defined and considered as part of the change?
- Are appropriate resources provided to support the process of change?
- Have all the relevant risks been defined and considered,

Responsibilities and accountability of different groups should be also defined and understood in order to guide the change successfully.

7.4. ASSESSMENT AND APPROVAL OF THE PROPOSED CHANGE

The primary purpose of the safety assessment of the proposal is to confirm that an adequate level of safety will be maintained and not be degraded or compromised. A safety assessment has been carried out at the early stage, when establishing the strategy. This stage is the assessment of the completed proposal and should be carried out by the proposer or initiator before any independent assessment (see Section 7.5). The operating organization is responsible for establishing the method and the requirements for the assessment process to be carried out. The process itself should adopt a graded approach by marking the importance and safety impact of the change. The safety assessment shall be performed in steps that include assessment carried by initiator, nuclear safety specialist and later on by independent supervisors if needed. The results of the assessment shall be approved by competent staff.

7.4.1. Carrying out safety categorization of the change

The organizational changes should be categorized from the point of view of their impact on nuclear and radiation safety in accordance with national regulatory legal requirements. Account has to be taken of operational experience and safety indicators. The initial category proposal has to be made by initiator or delegated specialist from the responsible department.

How should changes be categorized?

- Define the categories for the potential risk:
  - Identify a small number of categories from major down to insignificant suitable for the operating environment.
- There may be modifications that are outside this process, if the change does not affect safety. Though all changes should be managed for business purposes, it must be recognized that not all changes have an impact on safety; other processes and controls may then be used.
- Identify the risks to safe operation and compare the risks with the categories.

— Categorization should be checked by a competent person. Training is needed to carry out categorization. Awareness training may be needed for other personnel.

Recommended safety categories:

— Category 1 — Major impact on safety that changes the declared base for licence conditions (e.g. technical specification, personnel qualification, safety analysis, safety report).
— Category 2 — Impact on safety with consequences that do not lead to licence change.
— Category 3 — Limited or no impact on safety.

Category 1 changes may require approval by the regulatory body.

The categorization of the change results in need of safety analysis and its scope and the level of approval. The categorization can be changed in any stage of the change assessment, including the regulatory requirements.

Incorrect categorization may result in the proposal having inadequate independent scrutiny, with a resultant increase in the risk to the nuclear safety or spending excessive resource on a lesser change — and the risk that the process may fall into disrepute.

Some examples are presented below.

EXAMPLE 1. THE UK NUCLEAR INDUSTRY CODE OF PRACTICE (NICOP) [13]

Category	Definition	Level of scrutiny	Approval level
A Major effect.	Change with a major nuclear safety impact: — Large scale downsizing or outsourcing of a significant nuclear safety function; — Change that affects the legal basis of the license.	— Independent assessment by SQEP. — Relevant committee for endorsement. — Regulator for agreement.	— Licensee director/board member.
B Significant effect.	Change with a significant nuclear safety impact: — Wide ranging change resulting in significant transfer of responsibilities and accountabilities.	— Independent assessment by SQEP. — Relevant committee for review. — Regulator for review	— Functional director.
C Minor effect.	Change with a minor nuclear safety impact. This includes: — Change within a business/ directorate with nuclear safety responsibilities; — Change that has a minor impact on the company's emergency response organization.	— Independent assessment by SQEP. — Relevant committee for review. — Line manager or MOC coordinator.	— Sponsoring manager.
D Insignificant effect	— Change with negligible nuclear safety impact. — Change in a function/department/ individual post with little or no impact on nuclear safety.	— Line manager or MOC coordinator. — Independent review of categorization (which can be retrospective)	— Line manager.

EXAMPLE 2. SE-ENEL — GRADING OF CHANGES

Consideration is given to grading within the 'management of change' process utilized by the SE-Enel utility.

All changes are classified according to their safety, business, and environmental and other significance to ensure that appropriate controls are established and implemented.

In addition, the grading also considers the consequences in case that the change would be inadequately conceived or implemented.

Thus all changes in Slovenske Elektrarne, their assessment and approval routes use a graded approach based on the potential impact on nuclear safety (significance of the change). The following levels of organizational changes are defined:

- Level A: An organizational change related to those activities that have a direct impact on nuclear safety.
- Level B: An organizational change includes activities defined by the organization that have a potential impact on nuclear safety.
- Level C: An organizational change (mostly of a fundamental nature), which is not characterized as A or B, but if implemented can influence, providing for activities with direct impact and/or potential impact on safety (e.g. activities related to human resources, management, services, or procurement management). For such change it has to be proved that it will not have an impact on safety, by preparing the detailed plan of implementation of the change (see implementation plan).
- Level D: An organizational change of a more complex nature without any impact on nuclear safety, which does not include any activity related to the use of nuclear energy in the utility.
- Level E: An organizational change without any impact on nuclear safety, so-called simple changes meeting the following conditions:
  - The change has an impact only on activities within one unit;
  - The change does not include a decrease in the number of employees;
  - The change does not have an impact on the IMS documentation.

Note: An example of a simple change can be creation of a position in accordance with the plan of personal costs, or transfer of a position within a unit without any impact on the activities of the unit.

7.4.2. Assessment of the safety consequences

This is the opportunity to assess the risks. The organization's procedures must require an assessment of the safety significance to be carried out and this assessment must be rigorous and systematic. It must include analyses for different operating states, conditions and operational occurrences.

The change and the safety considerations must be adequately documented using the organization's normal management system approach and take account of the level in the organizational structure. The safety criteria should be extracted from basic design criteria and the organizational policy using results from periodic safety review. Particular focus should be given to the human factor with regard to personnel qualification and ability to fulfill the given responsibility. The safety culture should be maintained.

There are several approaches that can be followed, but each depends on the risks being correctly identified. This includes not only the resources (numbers of people, competences etc.), but also the attitudes, beliefs and values of the personnel affected (the 'culture').

This example has been taken from the UK Nuclear Industry Code of Practice (NICOP) [13]:

Structure

- Organization design, e.g. number of layers and extent of control;
- Governance arrangements;
- Control and supervision arrangements.

Accountability

— Clarity of accountability;
— Use of consistent post/role titles;
— Roles and responsibilities;
— Potential conflict between roles and priorities.

Resources

— Method of resourcing, e.g. replacing licensee staff with service providers;
— Risk of interruption to supply;
— Number and retention of SQEP resources;
— Maintenance of core capability;
— Availability, e.g. shift changes, workload;
— Competence levels, including competencies required by remaining personnel;
— Resources to implement the change;
— Gaps in the structure, e.g. vacancies;
— Resources and competences needed for safe operation/shutdown/emergency;
— Response of the facility;
— Technical and operational supporting resource;
— Effects of multiple skills;
— Singleton (single person with a particular set of competences) post or hardly available expertise;
— Staff turnover;
— Increased dependence on service providers;
— Retention of internal combustion capability;
— Corporate memory.

Management system

— Safety culture.
— Impact on arrangements to implement the nuclear site LCs.
— Compliance arrangements, e.g. will the transfer of role holders to another department leave a gap?
— Would revision to procedures or instructions be needed to reflect the change?
— Management of service providers' arrangements.
— Impact on resource demands of changing procedures or instructions.
— Will important communication and working relationships be affected?

It is necessary to ensure that the people carrying out the assessment have suitable qualifications and competence to carry out this task. In some regulatory regimes this will require a specific authorization. Following the generation of the risk assessment, consideration should be given to testing the assessment to confirm that risks are as low as reasonably practicable. Other options may be applicable to different organizations. A suggested risk assessment check sheet taken from the UK NICOP [13] is given in Annex IV.

7.4.3. Review, assessment and approval (internally and externally)

The complete proposal will require scrutiny before implementation. This step is the assessment of the *whole proposal* and consideration of whether the enablers actually address the risks adequately — and confirm the grading of the change. The extent and depth of the assessment will depend on the category or grade of the change. Lower category (minor) changes may only require review at the departmental level, but higher category changes that will introduce significant risk to nuclear safety should be subject to top level internal scrutiny, e.g. independent assessment and nuclear safety committee review followed by submission to the regulator for notification or agreement. Examples are given in Section 7.4.1 and a working example is given in Table 2 of Annex IV.

Assessment

It is also good practice for a nuclear safety committee and other internal assessment functions to conduct retrospective reviews of the totality of organizational changes. Organizations may choose to have a dedicated MOC committee(s) instead of using an existing nuclear safety committee.

How this is done

- Authors are encouraged to consult internal assessors at an early stage in the process; this helps with categorization and supports the preparation of the document.
- The assessor should review the proposed change to confirm (at a minimum):
  - That the basis is sound;
  - The risks have been correctly identified and assessed;
  - The enablers adequately address the risks.
- The assessor must consult the stakeholders regarding specific points beyond their direct competence.
- Once satisfied that the proposal is acceptable, the assessor should prepare a report recommending approval to the nominated body or individual. The level of authorization within the organization may depend upon the category of the change. *It is not normally the responsibility of the independent assessor to approve a proposal rather they may recommend approval and/or impose restrictions, requirements or caveats.*
- Depending upon the category of the change, the national regulator may be involved into the change process of the licensee under its jurisdiction.

A working example of an assessment process is given in Annex V (British Energy/EDF Procedure SRD/PROC/IN/006).

Approval

Approval of the proposal is the final stage before implementation. It is a separate stage to assessment and will be subject to the organization's procedures. No change may begin until it has been formally approved by the appropriate person, and any appropriate regulatory requirements have been met.

7.5. IMPLEMENTATION OF THE CHANGE

The next steps are applicable to any significant organizational change and are focused on aspects that have shown to be critical for success, as they relate to nuclear safety. Obviously, every change has needs in terms of resources, training, administrative acceptance, etc., but organizations normally have experience in dealing with these issues; consequently, this document especially takes account of human aspects. Any training requirements arising from the change must be identified in the implementation plan. Some of the steps may not need to be part of the implementation plan — in some organizations they are built into the management and safety cultures.

7.5.1. Preparation of a detailed implementation plan with resources, milestones and a review process

An implementation plan is essential. For simple changes, it may be sufficient to include enough detail in the change proposal. More significant and complex changes may need a more detailed, extensive and separate document.

The implementation plan will consist of a series of steps and instructions to facilitate the change. This must include all of the actions required after approval, but it may also suit some organizations to include the drafting, risk assessment, review and assessment steps which are necessary before approval is achieved. However, steps such as briefing, training or document preparation may need to be started before the change is actually enacted and specific enablers that will be necessary to facilitate the change. The plan must take account of the many factors that are needed to achieve a successful change and must refer to and align with the basic plan included in the proposal. It must identify timescales, milestones and any hold points.

At the purely functional level it is necessary to, as a minimum:

— Consult and brief the individuals affected;
— Prepare the detailed organizational charts (before, after and for any intermediate steps);
— Update or replace procedures and department manuals;
— Address any training, coaching or other facilitation needs;
— Retitle roles and/or realign training and competence records;
— Communications.

It may be useful to represent this as a project Gantt chart in addition to the detailed descriptive document. The process requires the identification of risks to nuclear safety; the proposal must identify the risks and one or more enabler to eliminate or mitigate each risk and the implementation plan must include a step or action for each of the enablers.

7.5.2. Establishing a sense of urgency

For any change to happen, it is important that the entire organization and other stakeholders understand the need and support it. It is necessary to explain the reasons for the change and the urgency (the need for change now) at both the rational and emotional levels, and a frank and deep dialogue is required about why the change is needed. This should develop the imperative for change; open discussion and dialogue will encourage a sustainable process — a sense of urgency will arise from strong perception of the need.

This can be achieved by:

— Developing scenarios showing what may happen if the change is successful or unsuccessful.
— Developing arguments for why the change is an opportunity to be grasped or the threats of being unsuccessful.
— Using any opportunity, small events, peer meetings or the experience of other plants to stress the importance of accomplishing the change successfully.
— Setting up honest discussions with people cascading down to all levels to get them talking, thinking and providing feedback about their support or concerns and fears.
— Enlisting support from key stakeholders (in addition to the internal stakeholders, it is important to deal with external stakeholders including any external regulatory body) to show the organization that there exist a shared understanding and commitment to the change.

It has been suggested that successful change depends on 'buy in' from at least 75% of the managers in the company; the effort required for this step should not be underestimated or undervalued. Insufficient attention will result in the need for extra time and resources.

7.5.3. Aligning the leadership team

This is a development of the step 'Define the guiding coalition' (Section 7.3.4). The leadership team is a key stakeholder in any change process and must be aligned with the need to change and with the chosen end point before the change can be sold to other stakeholders. This is part of the process of building the coalition to drive through the change. Once this has been achieved, the change coalition can be built. This coalition will typically consist of members of the management and leadership team plus 'experts' or others with suitable influence within the organization. The team must also include personnel who understand the impact of any change on nuclear safety as well as occupational safety.

To align the team:

— Use the guiding coalition to steer the leadership and staff towards the successful completion of the change.
— Influence members of the team from different levels and work areas within the organization if the new culture is to be made permanent.
— Include input from different team members so that the impact of the change is accounted for.

— Include members from each of the work areas affected to achieve 'buy-in'. Any weak areas must be identified.
— Stress the areas that will convince on an emotional as well as technical level. Rational alignment is not enough — it is necessary to obtain emotional buy-in if morale or other cultural aspects are not to be affected: 'hearts and minds'.
— Align the affected personnel, groups or departments. Effectively, the team grows to include all affected personnel.

It is important to recognize that there are different (business) cultures within any organization; communication will require certain skills beyond what is normally expected of a manager. The leadership team will require cross-cultural skills to cover this, which should not be restricted to managers.

The dangers of inadequate alignment include the risk of confusion, missed steps or incomplete understanding of the risks to safety.

7.5.4. Communicating the change vision

Communications relating to any change are not one way. There must be a true dialogue that produces a shared understanding of the vision and the stakeholders need to become engaged with it.

Communicating the vision: For any change to be effective and to become embedded, all of the stakeholders must be included in the dialogue, particularly those directly affected. This can be accomplished by:

— Talking about the vision at every opportunity.
— Explaining clearly the objective of the change.
— Using every existing communication channel, forum and briefing opportunity.
— Making a clear initial announcement.
— Keeping awareness high by:
  • Reinforcing the use of stage announcements and milestones;
  • Including daily updates and status reports;
  • Including routine messages, agenda items, etc.;
  • Including personal objectives.
— Listening for feedback and acknowledge and openly addressing concerns and anxieties.

It is also important that the leadership team 'walks the talk'. It must be a true dialogue. The dangers of inadequate communications are not only lack of awareness, but also that there will be inadequate engagement, misinterpretation of the vision, failure to fully implement parts of the change and damage to morale. It will fail, or at least fail to achieve the full benefit, and safety could be compromised.

7.5.5. Removing obstacles to improving nuclear safety

There will always be obstacles to a change and obstacles to improving nuclear safety in the existing arrangement. They may share certain features. The former must be addressed to complete the change, the latter so that the current level of safety and safety culture can be maintained or improved.

Obstacles have to be identified and addressed by:

— Recognizing that some obstacles may have been deliberately introduced to improve safety or safety culture in the past to avoid nuclear safety being compromised. The reasons for these may not be immediately obvious but the obstacles may be embedded.
— Considering if there are any features of the current arrangements that could obstruct the change, such as organizational structure, existing job descriptions or training requirements, appraisal, performance or compensation systems.
— Recognizing and rewarding people contributing to the success of the change

— Recognizing resistance to change, both deliberate and passive. Those resisting change must be treated with respect. Consider if they would respond to positive encouragement. Find the root cause of the resistance, and if consensus cannot be achieved then an alternative way forward may be needed.
— Taking action quickly to remove barriers once they are recognized.

The danger is that either the obstacle is used to demonstrate that the change cannot or will not be effective, or the resistance will delay the change unexpectedly.

7.5.6. Generating short term wins

Any progress encourages. People can be motivated by seeing progress and improvement and this will help maintain momentum.

Short term wins can be generated by the following:

— Short term wins are not always deliberately created, but may be tasks identified when the plan is first reviewed.
— Identifying tasks at the early stages of the plan — usually items with small targets — and ensuring that these are completed on time.
— Identifying any tasks in the plan that do not need substantial groundwork and complete them as early as possible. This has the added benefit of removing possible distractions later.
— Dividing larger or more onerous items into manageable chunks, and showing appreciation when early steps and milestones are achieved, e.g. rewriting the management system may be divided into steps for each type of document.
— Paying attention to results that can be celebrated, even if they were not originally identified as milestones.
— Recognizing participation and results as part of the formal appraisal process for individuals and accountability for departments.

Personnel will become demotivated or lose interest if there is little or no visible progress, the end of the change seems too far away, or their contribution is only a small part of the change. Acknowledge and appreciate the achievement when milestones are achieved and share with team so that everybody feels part of the success.

7.5.7. Consolidating gains and driving more change

Change requires continuous attention and successes must be continuously embedded. This will consolidate the changes achieved and stimulate the further changes in the project. To build on the gains achieved:

— Set goals which will build on the momentum achieved.
— Use the credibility that success brings to tackle the bigger and more difficult parts of the change.
— Identify what has gone well and any difficulties, and feed these back into the project; build on the momentum — continuous improvement — be open about successes and difficulties.
— Bring new people to the change coalition; this will make the project more inclusive and introduce new ideas; the team may be galvanized into action.

Change projects may fail if completion is declared too early. The infrastructure may be complete, but the new culture may not be accepted or embedded.

7.5.8. Measuring the change process and defining some adaptations

In order to ensure that a change is complete, that progress is being made, it is necessary to be able to measure the progress of the project, together with the quality of the results.

Progress can be measured by:

— Setting milestones and measuring the milestones against the plan, tasks and milestones achieved. However, it must be recognized that quality is as important as, or even more important than, quantity.
— Speaking to people, since qualitative measures are valuable. How the responses change over time should be noted.
— Using anonymous surveys (particularly for bigger changes).
— Defining numerical indicators in the proposal and seeking agreement on them with the senior manager responsible for the change. It should borne in mind that indicators are useful but may be misleading, so special attention should be paid to define them and ensure that they are specific, measurable, accepted, reasonable and timely (SMART). Since indicators can be misleading they must be supported by other measures.
— Observing behaviour, looking for acceptance, and observing whether the culture changed. For example:
  • Are new processes being followed?
  • Have attitudes changed?
— Observing whether the plan been changed to account for challenges, opportunities and delays arising from intermediate results.

If progress is not measured, achievement is not recognized, delays may not be apparent and the need or opportunity to employ a countermeasure may be missed. If the situation has changed and progress has not been measured, then the plan may no longer be valid; plans and timescales may need adjustment or even need radical overhaul.

7.5.9. Embedding new approaches in the culture

The change needs to become sustainable. The benefits need to be realized to the extent possible, but this might take some time. To make the change sustainable, it is necessary to embed the change and make it become 'normal business' as quickly as possible.

Change can be embedded by:

— Showing how the change/new process has improved performance, productivity, safety or some other obvious feature of the business;
— Reinforcing good behaviour that aligns with the new arrangement;
— Including the change achievements and ideals in briefing materials, particularly in briefing and training for new staff;
— Recognizing members of the change coalition for their contribution and, where possible, identifying a specific example that has benefited the business or, specifically, safety;
— Ensuring that leaders continue to support the change. Succession planning is key, the legacy of the change must not be lost.

There are a number of risks if the change is not properly embedded as normal business. Old approaches may continue to be used or may reappear, even though they no longer fit with the business. Unanticipated nuclear safety risks may materialize. The hard won benefits may be lost. The credibility of the change process may be damaged and this may affect nuclear safety directly or in the long term.

7.5.10. Assessing the change: Self-assessment and feedback after the change

The following questions must be posed at the end of the change process as the final step of the implementation plan, or as a post-implementation review (PIR) or self-assessment. Has the change been completed? Has the change been effective? What went well? What went wrong? What lessons have been learned? What benefits do people really appreciate? This assessment needs to be led by a senior officer who is in a position to challenge managers, should be defined in advance in the proposal, and should result in a formal report which is fed back into the normal management process. Such a review should not normally be carried out until perhaps six, twelve months or longer after the notional end point to determine if the change has become fully effective and properly embedded, depending on the extent of the change and the regulatory regime and company practice. Such a PIR may be a formal requirement

The result of the change can be assessed by:

— Recognizing that this is part of the measuring process (see Section 7.3.2). Use the measures that have been used throughout the change.
— Checking whether the objectives stated in the proposal have been achieved.
— Checking whether all of the enablers have been completed.
— Assessing whether the revised roles and responsibilities are clear.
— Checking that any additional or unexpected issues or training needs have been identified and managed.
— Checking that the communication plan is effective by:
  • Speaking with the people affected.
  • Checking that individuals were sufficiently aware and consulted.
— Assessing whether it was necessary to activate any contingency plans.
— Ensuring that good work, and the people responsible for it, has been recognized.
— Examining whether there been any impact on morale that could adversely affect the business or financial stability, the safety culture or nuclear safety.
— Examining what lessons can be learned from this change that might be useful for future changes.

Lessons learned from organizational changes should be documented and, where appropriate, fed into the experience, feedback and learning functions. Use any company corrective action programs. If the review finds that the change was poorly conceived or implemented, additional actions may be required to address any issues identified. This may prompt a further review, further MOC actions or even a reversal of the change. A working example of a formal PIR process is included in Annex IV.

8. RELATIONSHIP BETWEEN THE STEPS TO IMPLEMENT A CHANGE AND THEORETICAL ASPECTS

This section helps link the theoretical and background considerations to the practical steps. Suggested theoretical references are given for each of these steps.

Section/steps of the process	Related sections discussing theoretical aspects
7.1. Introduction	Foreword. 1. Introduction.
7.2. Managing organizational changes	2. Managing organizational change. 2.1. Model for managing organizational change.
7.3. Identification of the proposed changes	
7.3.1. Describe a need for change	2.3. Determinants/forces initiating change. 3.4. Anticipating organizational change. 3.5. Six types of 'evolutions' to consider. 4.2.4. Assessment of the competence and capability of licensees.
7.3.2. Define parameters to assess success of the proposed change.	2.5.4. Factors that determine success. Annex I–3. Appreciative inquiry.

Section/steps of the process	Related sections discussing theoretical aspects
7.3.3. Identify risks, enablers and countermeasures.	2.7.2. Why is organizational change difficult to accomplish? 2.7.3. Why do organizational changes fail? 3.2. Practices to counter safety problems during organizational change. 4.1. Organizational culture. 4.2.6. Mutual understanding through communication.
7.3.4. Define the guiding coalition.	2.4. Organizational initiator. 5.2. Leadership and management roles in change. 5.3. Training and coaching. Annex I–5. Leadership.
7.3.5. Develop a vision and strategy to implement the change.	I–1.5. Establishing an organizational culture that supports projects. Annex I–2. Leadership. Annex I–3. Appreciative inquiry. Annex I–6. Providing recognition and reward.
7.3.6. Prepare a formal proposal.	2.5. Intervention strategies. 2.6.1. Change process. 2.8. Transition management. 2.8.1. Activity planning. 2.8.3. Intervention methods. 2.8.4. Transition management structures. Annex I–3. Appreciative inquiry. Annex I–4. Levels of change.
7.4. Assessment and approval of the proposed change.	
7.4.1. Carry out safety categorization of the change.	
7.4.2. Assess the safety consequences.	2.2. Potential impact on safety of organizational change. 3.1. Potential safety problems caused by organizational change. 4.1. Organizational culture.
7.4.3. Review, assessment and approval (internally and externally).	4.2.1. Regulatory role in changing environment. 4.2.3. Regulatory tools. 4.2.4. Assessment of the competence and capability of licensees. 4.2.6. Mutual understanding through communication.
7.5. Implementation of the change.	
7.5.1. Prepare a detailed implementation plan with resources, milestones and review process.	2.6. Implementation. 2.7. Implementation tactics. 2.9. Integrated programme for organizational change. 5.1. Role of teams in organizational change. Annex I–5. Establishing an organizational culture that supports Annex I–5. Assessing whether an organization is ready for change.

Section/steps of the process	Related sections discussing theoretical aspects
7.5.2. Establish a sense of urgency.	5.1.1. Essential role of participation. 5.2. Leadership and management roles in change. Annex I–2. Leadership. Annex I–3. Appreciative inquiry. Annex I–6. Providing recognition and reward.
7.5.3. Align the leadership team (a development of the step 'Form a Powerful Coalition').	4.1. Organizational culture. 5.1.1. Essential role of participation. 5.2. Leadership and management roles in change. 5.3. Training and coaching. Annex I–2. Leadership. Annex I–6. Providing recognition and reward.
7.5.4. Communicate the change vision.	2.7.1. Communication to help lower resistance. 4.3.1. Communication fundamentals. 4.3.2. Avoiding common pitfalls. 4.3.3. Phases of the communication process. 4.3.4. Cross-cultural communication. Annex I–2. Leadership.
7.5.5. Remove obstacles to improving nuclear safety.	4.1. Organizational culture. 5.2. Leadership and management roles in change. Annex I–1. Role of organizational structure. Annex I–2. Leadership.
7.5.6. Generate short term wins.	2.8.2. Where to intervene first. 4.3.1. Communication fundamentals. 4.3.2. Avoiding common pitfalls. 4.3.3. Phases of the communication process. 4.3.4. Cross-cultural communication. 5.1.1. Essential role of participation. Annex I–6. Providing recognition and reward.
7.5.7. Consolidate gains and drive more change.	4.3.1. Communication fundamentals. 4.3.2. Avoiding common pitfalls. 4.3.3. Phases of the communication process. 4.3.4. Cross-cultural communication. 5.1.1. Essential role of participation. 5.2. Leadership and management roles in change. Annex I–2. Leadership. Annex I–3. Appreciative inquiry. Annex I–6. Providing recognition and reward.

Section/steps of the process	Related sections discussing theoretical aspects
7.5.8. Measure the change process and define some adaptations.	4.2.5. Measuring, monitoring and inspecting by regulators.
7.5.9. Embed new approaches in the culture.	4.1. Organizational culture. 5.1.1. Essential role of participation. Annex I–2. Leadership. Annex I–6. Providing recognition and reward.
5.10. Assess the change: self-assessment and feedback after the change	Annex I–4. Levels of change. Annex I–6. Providing recognition and reward.

REFERENCES

[1] INTERNATIONAL ATOMIC ENERGY AGENCY, The Management System for Facilities and Activities, IAEA Safety Standards Series No. GS-R-3, IAEA, Vienna (2006).

[2] INTERNATIONAL ATOMIC ENERGY AGENCY, Application of the Management System for Facilities and Activities, IAEA Safety Standards Series No. GS-G-3.1, IAEA, Vienna (2006).

[3] INTERNATIONAL ATOMIC ENERGY AGENCY, The Management System for Nuclear Installations, IAEA Safety Standards Series No. GS-G-3.5, IAEA, Vienna (2009).

[4] INTERNATIONAL ATOMIC ENERGY AGENCY, Managing Change in Nuclear Utilities, IAEA-TECDOC-1226, IAEA, Vienna (2001).

[5] INTERNATIONAL ATOMIC ENERGY AGENCY, Management of Continual Improvement for Facilities and Activities: A Structured Approach, IAEA-TECDOC-1491, IAEA, Vienna (2006).

[6] SENGE, P.M., The Fifth Discipline: The Art and Practice of the Learning Organization, Currency/Doubleday, New York (1990).

[7] SENGE, P.M, et al., The Dance of Change: The Challenges to Sustaining Momentum in Learning Organizations, Currency/Doubleday, New York (1999).

[8] HARVARD BUSINESS SCHOOL PRESS (Ed.), Managing Change to Reduce Resistance, Harvard Business School Press, Cambridge, MA (2005).

[9] TOKYO ELECTRIC POWER COMPANY, Fukushima Nuclear Accident Analysis Report, TEPCO, Tokyo (2012).

[10] INTERNATIONAL ATOMIC ENERGY AGENCY, IAEA Action Plan on Nuclear Safety, IAEA, Vienna (2011).

[11] INTERNATIONAL ATOMIC ENERGY AGENCY, IAEA International Fact Finding Expert Mission of the Fukushima Daiichi NPP Accident Following the Great East Japan Earthquake and Tsunami, Mission Report: The Great East Japan Earthquake Expert Mission, IAEA, Vienna (2011).

[12] KOTTER, J.P., Leading Change, Harvard Business School Press, Cambridge, MA (1996).

[13] SAFETY DIRECTORS FORUM, Nuclear Baseline and the Management of Organisational Change, Nuclear Industry Code of Practice (NICOP), United Kingdom (2010).

BIBLIOGRAPHY

BECKHARD, R., HARRIS, R.T., Organizational Transitions, Addison-Wesley, Boston (1987).

GALPIN, T.J., The Human Side of Change, Jossey-Bass, San Francisco (1996).

HALL, J., HAMMOND, S., What is Appreciative Inquiry?, Thinbook Publishing, Bend, OR (2000).

INTERNATIONAL ATOMIC ENERGY AGENCY, Developing Safety Culture in Nuclear Activities: Practical Suggestions to Assist Progress, Safety Reports Series No. 11, IAEA, Vienna (1998).

INTERNATIONAL ATOMIC ENERGY AGENCY, Knowledge Management for Nuclear Industry Operating Organizations, IAEA-TECDOC-1510, IAEA, Vienna (2006).

INTERNATIONAL ATOMIC ENERGY AGENCY, Operational Safety Performance Indicators for Utilities, IAEA-TECDOC-1141, IAEA, Vienna (2000).

INTERNATIONAL ATOMIC ENERGY AGENCY, Risk Management: A Tool for Improving Utility Performance, IAEA-TECDOC-1209, IAEA, Vienna (2001).

INTERNATIONAL ATOMIC ENERGY AGENCY, Safety Culture in Nuclear Installations, IAEA-TECDOC-1329, IAEA, Vienna (2002).

INTERNATIONAL ATOMIC ENERGY AGENCY, Self-Assessment of Operational Safety for Utilities, IAEA-TECDOC-1125, IAEA, Vienna (1999).

KILMANN, R.H., Managing beyond the Quick Fix, Jossey-Bass, San Francisco (1989).

MATURANA, H., VARELA, F., The Tree of Knowledge: The Biological Roots of Human Understanding (Revised Edition), Shambhala Publications, Boston (1992).

NUTT, P.C., Tactics of Implementation, Acad. Manage. J. (1986).

PENDLEBURY, J., GROUARD, B., MESTON, F., Successful Change Management, Wiley, Chichester (1998).

SAFETY DIRECTORS FORUM, Nuclear Industry Code of Practice (NICOP) (UK) — Nuclear Baseline and the Management of Organisational Change (2010).

STEWART, T.A., Rate your readiness to change, Fortune (7 February 1994).

TAYLOR, F.W., The Principles of Scientific Management, Harper, New York and London (1911).

CONTENTS OF THE COMPANION CD-ROM

CONTRIBUTORS TO DRAFTING AND REVIEW

Bannister, S.	BNFL, United Kingdom
Boogaard, J.P.	International Atomic Energy Agency
Brandt, T.	Nuclear Competence Centre, Finland
Clark, R.	Consultant, United States of America
Cook, J.	Scientech, United States of America
Dahlgren Persson, K.	International Atomic Energy Agency
Drábová, D.	SUJB, Czech Republic
Durham, L.	Sterling Learning Services, Inc., United States of America
Forshaw, M.	EDF Energy, United Kingdom
Gardelliano, S.	International Consultancy & Learning Facilitation, Italy
Kawamura, S.	TEPCO, Japan
Kulkova, H.	SUJB, Czech Republic
Majola, J.	International Atomic Energy Agency
Merry, M.	Safety culture consultant, United Kingdom
Molloy, B.	International Atomic Energy Agency
Nichols, R.	International Atomic Energy Agency
Pagannone, B.	International Atomic Energy Agency
Pasztory, Z.	Paks Nuclear Power Plant, Hungary
Patten, J.	International Atomic Energy Agency
Redman, N.	Amethyst Management Ltd, United Kingdom
Takekuro, I.	TEPCO, Japan
Williamson, B.	Interlogic, Inc., United States of America
Villadoniga, J.	Tecnatom, Spain
Vincze, P.	International Atomic Energy Agency
Zlatnansky, J.	Slovenske Elektrarne, Slovakia

Consultants meetings

Vienna, Austria: 9–11 November 2004; 6–8 September 2005; 10–14 October 2011; 16–20 January 2012

Technical meeting

Vienna, Austria: 17–20 May 2011

Structure of the IAEA Nuclear Energy Series

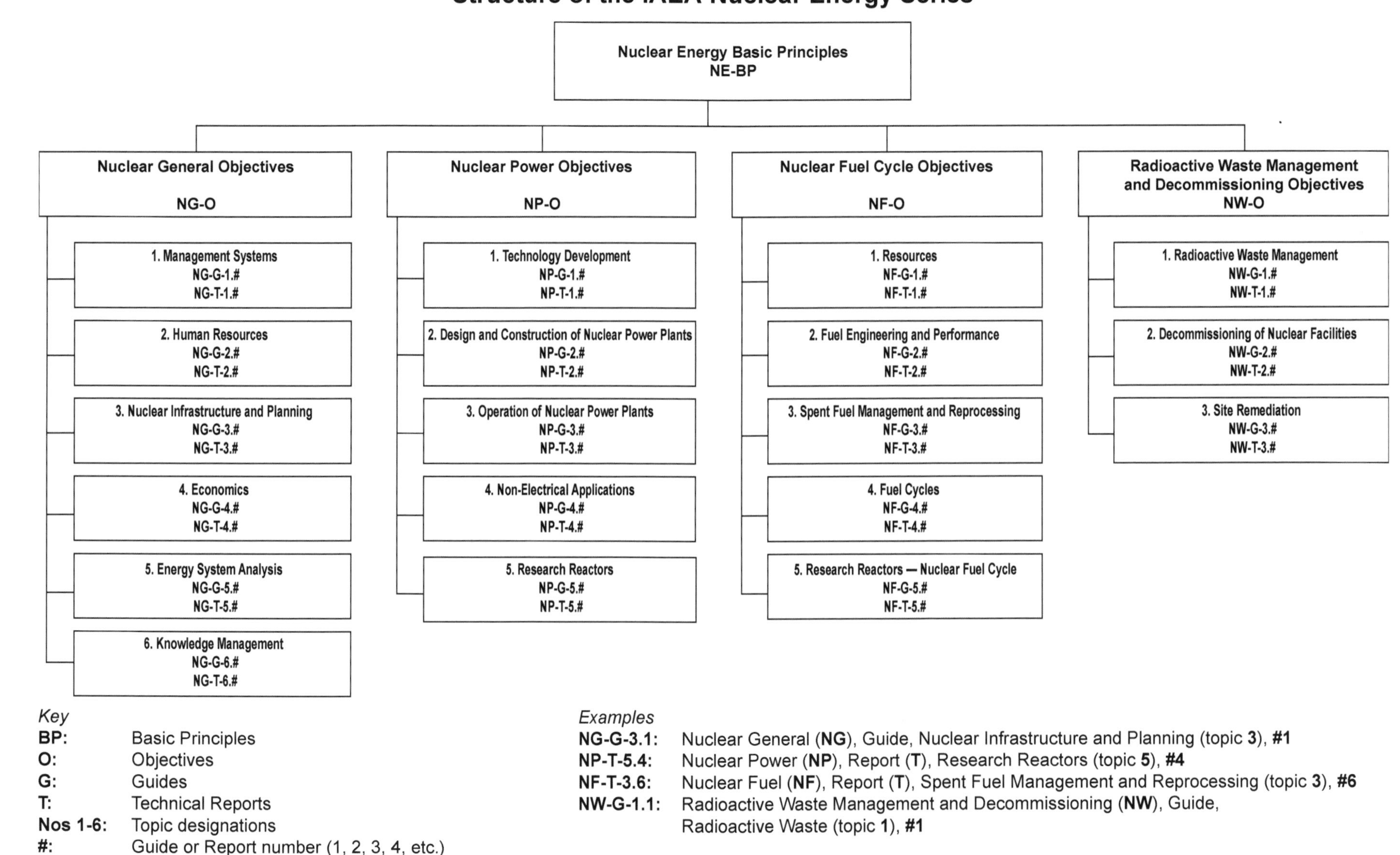

Key

BP: Basic Principles
O: Objectives
G: Guides
T: Technical Reports
Nos 1-6: Topic designations
#: Guide or Report number (1, 2, 3, 4, etc.)

Examples

NG-G-3.1: Nuclear General (**NG**), Guide, Nuclear Infrastructure and Planning (topic **3**), **#1**
NP-T-5.4: Nuclear Power (**NP**), Report (**T**), Research Reactors (topic **5**), **#4**
NF-T-3.6: Nuclear Fuel (**NF**), Report (**T**), Spent Fuel Management and Reprocessing (topic **3**), **#6**
NW-G-1.1: Radioactive Waste Management and Decommissioning (**NW**), Guide, Radioactive Waste (topic **1**), **#1**

No. 23

ORDERING LOCALLY

In the following countries, IAEA priced publications may be purchased from the sources listed below or from major local booksellers.

Orders for unpriced publications should be made directly to the IAEA. The contact details are given at the end of this list.

AUSTRALIA
DA Information Services
648 Whitehorse Road, Mitcham, VIC 3132, AUSTRALIA
Telephone: +61 3 9210 7777 • Fax: +61 3 9210 7788
Email: books@dadirect.com.au • Web site: http://www.dadirect.com.au

BELGIUM
Jean de Lannoy
Avenue du Roi 202, 1190 Brussels, BELGIUM
Telephone: +32 2 5384 308 • Fax: +32 2 5380 841
Email: jean.de.lannoy@euronet.be • Web site: http://www.jean-de-lannoy.be

CANADA
Renouf Publishing Co. Ltd.
5369 Canotek Road, Ottawa, ON K1J 9J3, CANADA
Telephone: +1 613 745 2665 • Fax: +1 643 745 7660
Email: order@renoufbooks.com • Web site: http://www.renoufbooks.com

Bernan Associates
4501 Forbes Blvd., Suite 200, Lanham, MD 20706-4391, USA
Telephone: +1 800 865 3457 • Fax: +1 800 865 3450
Email: orders@bernan.com • Web site: http://www.bernan.com

CZECH REPUBLIC
Suweco CZ, spol. S.r.o.
Klecakova 347, 180 21 Prague 9, CZECH REPUBLIC
Telephone: +420 242 459 202 • Fax: +420 242 459 203
Email: nakup@suweco.cz • Web site: http://www.suweco.cz

FINLAND
Akateeminen Kirjakauppa
PO Box 128 (Keskuskatu 1), 00101 Helsinki, FINLAND
Telephone: +358 9 121 41 • Fax: +358 9 121 4450
Email: akatilaus@akateeminen.com • Web site: http://www.akateeminen.com

FRANCE
Form-Edit
5 rue Janssen, PO Box 25, 75921 Paris CEDEX, FRANCE
Telephone: +33 1 42 01 49 49 • Fax: +33 1 42 01 90 90
Email: fabien.boucard@formedit.fr • Web site: http://www.formedit.fr

Lavoisier SAS
14 rue de Provigny, 94236 Cachan CEDEX, FRANCE
Telephone: +33 1 47 40 67 00 • Fax: +33 1 47 40 67 02
Email: livres@lavoisier.fr • Web site: http://www.lavoisier.fr

L'Appel du livre
99 rue de Charonne, 75011 Paris, FRANCE
Telephone: +33 1 43 07 50 80 • Fax: +33 1 43 07 50 80
Email: livres@appeldulivre.fr • Web site: http://www.appeldulivre.fr

GERMANY
Goethe Buchhandlung Teubig GmbH
Schweitzer Fachinformationen
Willstätterstrasse 15, 40549 Düsseldorf, GERMANY
Telephone: +49 (0) 211 49 8740 • Fax: +49 (0) 211 49 87428
Email: s.dehaan@schweitzer-online.de • Web site: http://www.goethebuch.de

HUNGARY
Librotade Ltd., Book Import
PF 126, 1656 Budapest, HUNGARY
Telephone: +36 1 257 7777 • Fax: +36 1 257 7472
Email: books@librotade.hu • Web site: http://www.librotade.hu

INDIA

Allied Publishers

1st Floor, Dubash House, 15, J.N. Heredi Marg, Ballard Estate, Mumbai 400001, INDIA
Telephone: +91 22 2261 7926/27 • Fax: +91 22 2261 7928
Email: alliedpl@vsnl.com • Web site: http://www.alliedpublishers.com

Bookwell

3/79 Nirankari, Delhi 110009, INDIA
Telephone: +91 11 2760 1283/4536
Email: bkwell@nde.vsnl.net.in • Web site: http://www.bookwellindia.com

ITALY

Libreria Scientifica "AEIOU"

Via Vincenzo Maria Coronelli 6, 20146 Milan, ITALY
Telephone: +39 02 48 95 45 52 • Fax: +39 02 48 95 45 48
Email: info@libreriaaeiou.eu • Web site: http://www.libreriaaeiou.eu

JAPAN

Maruzen Co., Ltd.

1-9-18 Kaigan, Minato-ku, Tokyo 105-0022, JAPAN
Telephone: +81 3 6367 6047 • Fax: +81 3 6367 6160
Email: journal@maruzen.co.jp • Web site: http://maruzen.co.jp

NETHERLANDS

Martinus Nijhoff International

Koraalrood 50, Postbus 1853, 2700 CZ Zoetermeer, NETHERLANDS
Telephone: +31 793 684 400 • Fax: +31 793 615 698
Email: info@nijhoff.nl • Web site: http://www.nijhoff.nl

Swets Information Services Ltd.

PO Box 26, 2300 AA Leiden
Dellaertweg 9b, 2316 WZ Leiden, NETHERLANDS
Telephone: +31 88 4679 387 • Fax: +31 88 4679 388
Email: tbeysens@nl.swets.com • Web site: http://www.swets.com

SLOVENIA

Cankarjeva Zalozba dd

Kopitarjeva 2, 1515 Ljubljana, SLOVENIA
Telephone: +386 1 432 31 44 • Fax: +386 1 230 14 35
Email: import.books@cankarjeva-z.si • Web site: http://www.mladinska.com/cankarjeva_zalozba

SPAIN

Diaz de Santos, S.A.

Librerias Bookshop • Departamento de pedidos
Calle Albasanz 2, esquina Hermanos Garcia Noblejas 21, 28037 Madrid, SPAIN
Telephone: +34 917 43 48 90 • Fax: +34 917 43 4023
Email: compras@diazdesantos.es • Web site: http://www.diazdesantos.es

UNITED KINGDOM

The Stationery Office Ltd. (TSO)

PO Box 29, Norwich, Norfolk, NR3 1PD, UNITED KINGDOM
Telephone: +44 870 600 5552
Email (orders): books.orders@tso.co.uk • (enquiries): book.enquiries@tso.co.uk • Web site: http://www.tso.co.uk

UNITED STATES OF AMERICA

Bernan Associates

4501 Forbes Blvd., Suite 200, Lanham, MD 20706-4391, USA
Telephone: +1 800 865 3457 • Fax: +1 800 865 3450
Email: orders@bernan.com • Web site: http://www.bernan.com

Renouf Publishing Co. Ltd.

812 Proctor Avenue, Ogdensburg, NY 13669, USA
Telephone: +1 888 551 7470 • Fax: +1 888 551 7471
Email: orders@renoufbooks.com • Web site: http://www.renoufbooks.com

United Nations

300 East 42nd Street, IN-919J, New York, NY 1001, USA
Telephone: +1 212 963 8302 • Fax: 1 212 963 3489
Email: publications@un.org • Web site: http://www.unp.un.org

Orders for both priced and unpriced publications may be addressed directly to:

IAEA Publishing Section, Marketing and Sales Unit, International Atomic Energy Agency
Vienna International Centre, PO Box 100, 1400 Vienna, Austria
Telephone: +43 1 2600 22529 or 22488 • Fax: +43 1 2600 29302
Email: sales.publications@iaea.org • Web site: http://www.iaea.org/books